Systems Biology

Topical Conference at the 2010 AIChE Annual Meeting

Salt Lake City, Utah, USA
7-12 November 2010

ISBN: 978-1-61782-141-7

Printed from e-media with permission by:

Curran Associates, Inc.
57 Morehouse Lane
Red Hook, NY 12571

CURRAN ASSOCIATES INC.
proceedings.com

Some format issues inherent in the e-media version may also appear in this print version.

Printed by Curran Associates, Inc. (2011)

For permission requests, please contact AIChE
at the address below.

AIChE
3 Park Avenue
New York, NY 10016-5991

Phone: (203) 702-7660
Fax: (203) 775-5177

www.aiche.org

Additional copies of this publication are available from:

Curran Associates, Inc.
57 Morehouse Lane
Red Hook, NY 12571 USA
Phone: 845-758-0400
Fax: 845-758-2634
Email: curran@proceedings.com
Web: www.proceedings.com

TABLE OF CONTENTS

(66a) A Glucose Valve for Pathway Engineering

Kevin V. Solomon, Tae Seok Moon and Kristala L. J. Prather, Chemical Engineering, Synthetic Biology Engineering Research Center (SynBERC), Massachusetts Institute of Technology, Cambridge, MA

In engineering metabolic pathways, branch points frequently arise that compete for pathway intermediates, thus limiting productivity. Common solutions of knocking out these competing pathway genes, however, are infeasible when the heterologous pathway competes directly with central metabolism. In this study, we describe the engineering of *E. coli* to create a platform capable of directing unphosphorylated glucose towards heterologous pathways. We have created a viable *E. coli* host that decouples glucose transport and phosphorylation allowing us to independently control the flux of glucose. Through the use of glucokinase (*glk*) antisense RNA constructs, we are able to inhibit *glk* activity by up to 25% and perturb central metabolism. Further studies showed that these constructs are able to boost specific productivity in a model heterologous glucose-consuming pathway. The use of antisense RNA in this manner allows for fine analog control creating a 'glucose valve' and novel optimization opportunities for pathway development. In this talk, I will highlight some of the successes and limitations of this technology as well as future areas of development.

See more of this Session: Synthetic Systems Biology I
See more of this Group/Topical: Topical A: Systems Biology

(66c) Design and Construction of a Protein Degradation Switch

S. Marjan Varedi K.[1], Daniel M. Janse[2], George M. Church[2] and Xiaoxia (Nina) Lin[1], (1)Chemical Engineering, University of Michigan, Ann Arbor, MI, (2)Genetics, Harvard Medical School, Boston, MA
1st and 2nd authors: Equal Contributions

Targeted protein degradation in eukaryotic cells requires recruitment of the target protein to proteasome which is usually triggered by ubiquitination of the target protein. It has recently been shown that ubiquitination is not necessary and localization to the proteasome is sufficient for degradation [1]. One intriguing way to bypass the ubiquitination step is artificial induction of target protein and proteasome dimerization. In this method, both the target protein and the proteasome are fused to particular protein domains and upon addition of a small molecule that can simultaneously bind to both domains, dimerization and consequently localization of the target protein to the proteasome occur. To implement this idea, we took advantage of a naturally occurring dimerizing molecule, rapamycin. We fused the Rpn10 subunit of Saccharomyces cerevisiae's proteasome with Fpr1, and on the other hand we fused the target protein, such as Ura3, with the Fpr1-rapamycin binding domain of Tor1. Upon adding rapamycin to the cell culture medium, Fpr1 and Tor1 domains come together and hence the target protein is localized to the proteasome and degrades. We have shown previously that the above approach is applicable for certain target proteins [2]. However, its efficiency varies substantially across different proteins and the method remains ineffective for proteins that are inherently difficult to degrade. To overcome this limitation, we propose to utilize multiple binding modules to speed up the degradation process. After confirmation from simulation studies that increasing the number of binding modules would lead to faster degradation, we constructed a series of differently tagged proteins to test our hypothesis. To study systematically the effect of the number of binding modules, we made use of a mutated version of Tor1 that cannot bind to rapamycin. Triple tags with different combinations of wild-type and mutated Tor1 domains (eight combinations in total) were constructed and fused to Ura3. The temporal profile of Ura3 in response to rapamycin has been measured in various constructs and the half-life of the Ura3 protein has been estimated accordingly. Our results verify that increasing the number of wild-type Tor1 domains from one to three on the Ura3 protein accelerates degradation in response to rapamycin significantly. The half-life is decreased from about 30 min in the single-Tor1 construct to about 10 min in the triple-Tor1 construct. This work highlights the design-orientated approach of synthetic biology in engineering molecular tools for switch-like and tunable control of protein degradation.

[1] Stankunas K, Crabtree GR, Exploiting protein destruction for constructive use, PNAS, 2007, 104 (28): 11511-11512.

[2] Janse DM, Crosas B, Finley D, Church GM, Localization to the proteasome is sufficient for

degradation, J Biol Chem, 2004, 279(20): 21415-21420.

See more of this Session: Synthetic Systems Biology I
See more of this Group/Topical: Topical A: Systems Biology

(66d) Modeling Eukaryotic Unfolded Protein Response and Its Role in Cellular Stress

Anirikh Chakrabarti and Jeffrey D. Varner, Chemical and Biomolecular Engineering, Cornell University, Ithaca, NY

Malfunctions in the folding state of critical proteins have been linked with cancer, diabetes and other diseases. Proteins requiring complex post-translational modification are processed in the Endoplasmic Reticulum (ER). Cells monitor protein folding by an inbuilt quality-control system involving both the ER and the Golgi apparatus. Incorrectly folded proteins are tagged for degradation or sent back through a refolding cycle. However, accumulation of incorrectly folded proteins can also trigger a cascade of events, termed the Unfolded Protein Response (UPR). In this study, we developed a mathematical model of UPR, which was composed of a system of ordinary differential equations. The model was assembled as a series of molecular modules integrated together to form the UPR network. Kinetic parameters for the model were estimated by comparing simulations with a family of training constraints. Eight sets of western blot experiments were used as the training constraints. Using a multi-objective thermal annealing scheme, we generated an ensemble of models consistent with the western blot constraints. Model output was compared to eleven sets of western blot studies to test the correctness of the model. Sensitivity analysis revealed that two particular modules, the UPR initiation and apoptosis modules, were the most sensitive structural elements of the model. Coupling analysis further highlighted the importance of key nodes e.g., the release of BiP (GRP78) from the ER stress transducers or the activation of Activating Transcription Factor 4 (ATF4). Taken together, we demonstrated that modeling could be used to understand key structural elements of UPR, despite model uncertainty. Understanding the architecture of this stress pathway could help us understand the design principles governing other stress related networks, e.g., hypoxia. Thus, while the current study was limited to UPR, the general strategy could be extended to other stress networks relevant to human health.

See more of this Session: Synthetic Systems Biology I
See more of this Group/Topical: Topical A: Systems Biology

(66e) Regulatory Role of Tandem Repeats Containing Transcription Factor Binding Sites

Tek Hyung Lee and Narendra Maheshri, Chemical Engineering, MIT, Cambridge, MA

Tandem repeats of DNA are widely prevalent in higher eukaryotic genomes. Because their growth and decay can occur after only a few generations, the lengths of these tracts are often hypervariable and vary between individuals. Here we explore the regulatory impact that tandem repeats of DNA that contain transcription factor (TF) binding sites can have on gene expression, using a synthetic biology approach in budding yeast. Not surprisingly, we find that the addition of arrays of repeated binding sites for a transcriptional activator competitively inhibit the normal TF-promoter interactions and reduce gene expression. More surprisingly, the repeat sites convert the linear dose-response of the promoter with respect to the TF into a more ultrasensitive response. This can be explained by a model whereby the repeat sites have 10-100 fold higher affinity to the TF compared with the promoter. We discuss the origins and implications of this difference and demonstrate how the changes in the repeat number of off-target binding sites can qualitatively affect gene regulation.

See more of this Session: Synthetic Systems Biology I
See more of this Group/Topical: Topical A: Systems Biology

(66f) Emulating Infections for Reprogramming Local Immune Responses

Nichole Daringer[1], Shirley Samuel-Abraham[1] and **Joshua N. Leonard**[2], (1)Chemical and Biological Engineering, Northwestern University, Evanston, IL, (2)Chemical and Biological Engineering & Robert H. Lurie Comprehensive Cancer Center, Northwestern University, Evanston, IL

Tumors and infectious pathogens evolve the capacity to manipulate host responses and induce local immune dysfunction. In particular, many primary and metastatic solid tumors induce profound immunosuppression, which impedes both natural immune clearance and interventions such as therapeutic cancer vaccination. Overcoming this barrier is widely considered a major obstacle to achieving safe and effective biological therapies for cancer. Recently, it was reported that such immunosuppression can be reversed by injecting small-molecule ligands for biosensor proteins of the innate immune system, called Toll-like Receptors (TLRs), into an established tumor. However, therapeutic utilization of this mechanism is complicated by the fact that (a) systemic administration of TLR ligands is potentially toxic and (b) local administration requires repeated (or continuous) injection of ligands into known tumor sites over an extended period of time. To overcome these limitations, we are employing the tools of synthetic biology to engineer programmable cellular devices that safely and specifically reproduce the potency of injected TLR ligands in a clinically viable fashion. In particular, we are developing genetically engineered macrophages (EMPs), that travel to tumor milieus, detect specified microenvironmental cues (e.g., elevated levels of cytokines such as IL-10 and TGF-beta), and induce potent pro-inflammatory responses (e.g., by inducing TLR-regulated gene networks). We are pursuing these goals by engineering receptor complexes (inputs), intracellular signaling and regulatory pathways (processors), and intercellular signaling mechanisms (outputs). Here we will describe the construction, characterization, and initial implementation of this novel platform for programming local immune function.

See more of this Session: Synthetic Systems Biology I
See more of this Group/Topical: Topical A: Systems Biology

(66g) Using Synthetic Biology to Engineer Syntrophic Microbial Communities for Bioprocess Applications

Ross P. Carlson and Hans Bernstein, Chemical and Biological Engineering, Montana State University, Bozeman, MT

A combination of synthetic biology and in silico network analysis was used to construct artificial microbial consortia comprised of engineered Escherichia coli strains. The design was based on biomimicry of key ecological roles found in stable, naturally occurring microbial consortia. In silico representations of the artificial communities were used to identify genetic modifications necessary for strategic community division of labor and to predict and interpret experimental interactions between strains. The artificial consortia were studied under both chemostat and biofilm growth conditions and demonstrated partitioning of electron donor and electron acceptor resources as designed. The artificial community metabolic interactions dampened chemostat oscillations associated with metabolic overflow production of inhibitory compounds like acetate highlighting ecological and bioprocess implications of syntrophhic consortia. The engineered community, when cultured as a biofilm, self-assembled into micron-scale spatial regions and represents a new tool for engineering multi-reaction bioprocess systems.

See more of this Session: Synthetic Systems Biology I
See more of this Group/Topical: Topical A: Systems Biology

(67a) Host Genetic Interactions During Viral Infection: A Systems Virology Approach to Lambda Phage Infection of E. Coli

Elsa Birch[1], Nathaniel Maynard[2] and Markus Covert[2], (1)Chemical Engineering, Stanford University, Stanford, CA, (2)Bioengineering, Stanford University, Stanford, CA

The complex infection process initiated by the viral genome arises from the interaction of the infectious agent with its host. Understanding this relatively small set of instructions as it interprets and co-opts its host's resources is crucial to addressing the problem of viral infection. Using the fundamental system of lambda phage and E. coli, we investigate the host-viral genetic interactions the disruption of which lead to a reduced infectivity. We gain additional insight into gene roles and interactions by observing population dynamics during infection. We gain additional insight into gene roles and interactions by observing population dynamics during infection, with particular focus on the role of central metabolism in the infection process.

See more of this Session: Systems and Synthetic Biology of Interacting Microorganisms
See more of this Group/Topical: Food, Pharmaceutical & Bioengineering Division

(67b) Metabolomics Reveals the Relationship Between Ketogulonigenium Vulgare and Bacillus Megaterium

Ying-jin Yuan, Department of Pharmaceutical Engineering, School of Chemical Engineering and Technology, Tianjin University, Tianjin, China

Ketogulonigenium vulgare and Bacillus megaterium were found to form a natural dual ecosystem and has been utilized to produce 2-keto-L-gulonic acid (2-KLG), precursor of vitamin C, for over 30 years. Understanding the metabolic interaction is critical to improving the vitamin C production. In this work, a comparative metabolomics analysis strategy was used to investigate the profiles of the molecular compositions of metabolites both in vivo and in vitro so as to elucidate the changes in metabolic interaction of this ecosystem. Significant changes in metabolites related to chemitaxis, central carbon metabolism and sporulation of Bacillus were explored.

See more of this Session: Systems and Synthetic Biology of Interacting Microorganisms
See more of this Group/Topical: Food, Pharmaceutical & Bioengineering Division

(67c) Community Modeling and the Design of Effective Uranium Bioremediation Strategy

Kai H. Zhuang[1], Eugene Ma[1], Melissa Barlett[2], Derek R. Lovley[2] and Radhakrishnan Mahadevan[1], (1)Chemical Engineering, University of Toronto, Toronto, ON, Canada, (2)Department of Microbiology, University of Massachusetts, Amherst, Amherst, MA

During uranium bioremediation at a contaminated site in Rifle, CO, there is an initial bloom of *Geobacter* accompanied by removal of U(VI) from the groundwater, followed by an increase of sulfate-reducing bacteria (SRBs) which are poor reducers of U(VI). Existing analysis suggests that there is competition between *Geobacter* and SRB, and when biologically usable and more energy efficient Fe(III) is depleted, the SRB compete more effectively for the substrate and able to grow using the available sulfate. However, the dynamics of the microbial competition at this site is not clearly understood, motivating the need for systems biology tools to understand the associated microbial ecology.

In order to understand the metabolic interactions in microbial communities, we have developed a dynamic genome-scale modeling approach denoted as the Dynamic Multi-species Metabolic Modeling framework. We have previously used this approach to model the community competition among Fe(III) reducers and had shown that the trade-off between growth rate and biomass yield under different environments was the decisive factor controlling the microbial competition at that site. In this work, we have used this framework to create a community model of *Geobacter* and SRBs. Batch simulations of community dynamics were used to examine the cause of the late onset of sulfate-reduction using different percentages of starting inoculants of *Geobacter* and SRB. These simulations showed that no matter what the starting conditions, the *Geobacter* were able to dominate the early period of the experiment with very little change in timing of the onset of sulfate-reduction. The field-scale simulation was able to accurately predict the field data (Figure 1), and suggests that the observed field data are due to growth rate dynamics and not due to thermodynamic competition.

Field-scale simulation of iron addition suggests that Fe(III) addition is a potential strategy for continuous uranium removal. We simulated the simultaneous addition of acetate and Fe(III) in both batch and field conditions using the community model. This showed that the addition of Fe(III) allowed *Geobacter* species to maintain its dominance over SRB. We then computationally optimized the rate of iron and acetate addition into the subsurface. We show that it is possible to maintain the uranium concentration below the environmental safety standards (Figure 2).

These simulations show that simultaneous addition of acetate and Fe(III) has the potential to be an effective uranium bioremediation strategy. They also show that computational modeling of microbial community is an important tool to design effective strategies for practical applications in biotechnology.

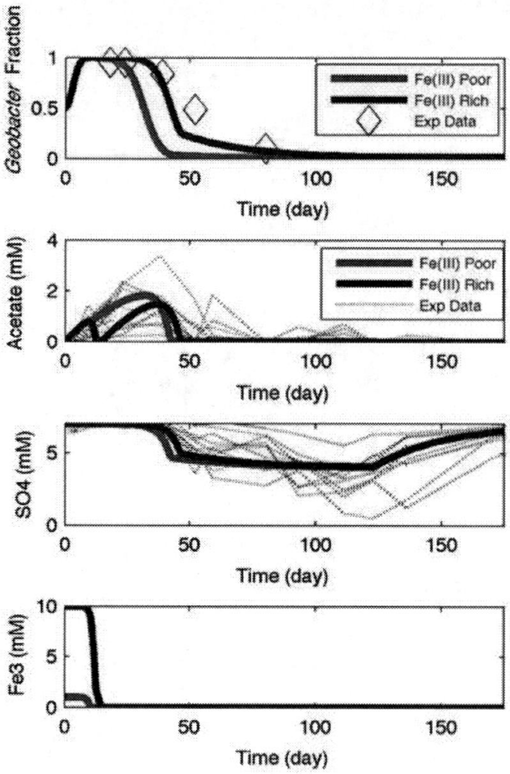

Figure 1. Field-scale simulation of Geobacter and SRB competition during acetate amendment.

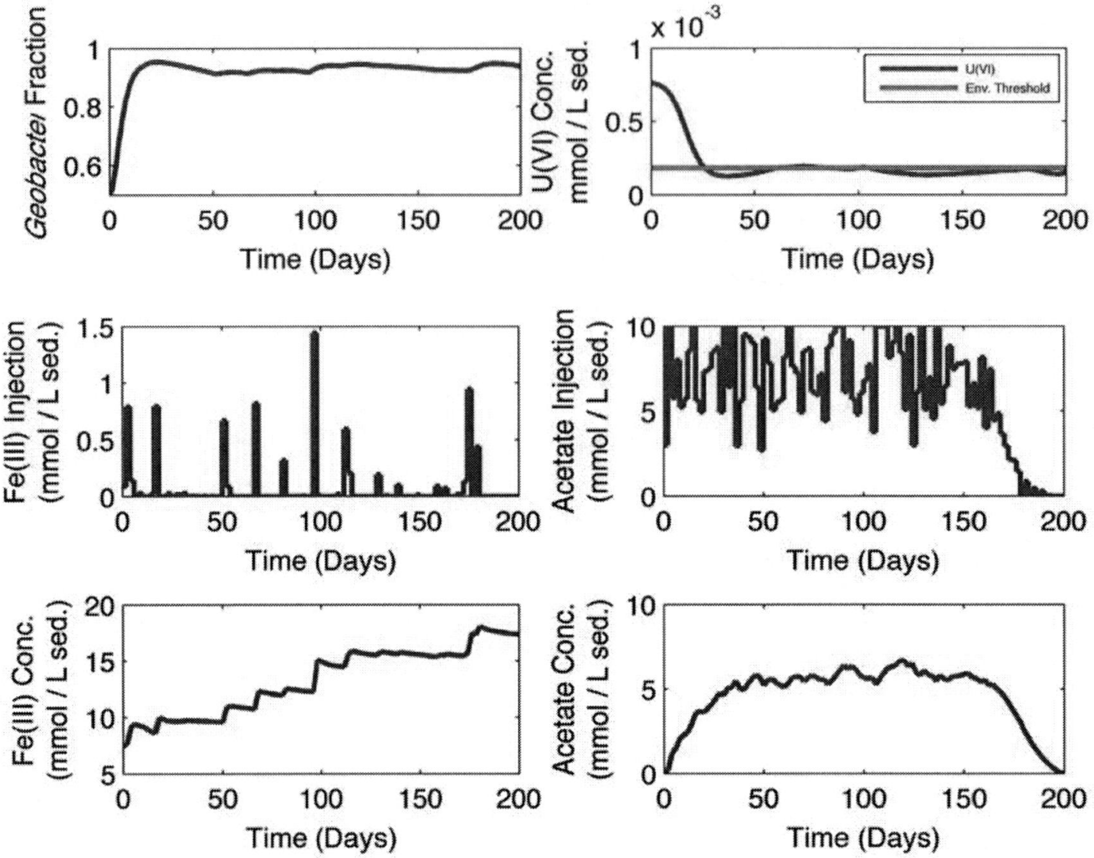

Figure 2. Computational Optimization of Acetate and Iron Addition Rates. Red line is the US Environmental Safety Level for Dissolved Uranium (VI)

See more of this Session: Systems and Synthetic Biology of Interacting Microorganisms
See more of this Group/Topical: Food, Pharmaceutical & Bioengineering Division

(67d) Quantitative Proteomic Analysis of Interactions in a Binary Bacterial Culture

Kenneth F. Reardon, Chemical and Biological Engineering, Colorado State University, Fort Collins, CO and Carla M. R. Lacerda, Department of Clinical Sciences, Colorado State University, Fort Collins, CO

Classic studies of biological interactions have relied on simple descriptions such as competition and antagonism, but these terms convey only an effect and do not provide any information on how the interaction affects the organisms at a molecular level. Experimental systems biology methods have enormous potential to provide information about biological interactions. Among the different high-throughput technologies, proteomic approaches offer some advantages when compared to DNA or RNA-based methods, including the ability to provide protein-level functional information. To evaluate the complexity of molecular-level responses associated with interactions, a quantitative proteomics study was conducted on a constructed binary bacterial culture. An iTRAQ shotgun proteomics workflow was used to compare the protein profiles of the Gram-negative *Pseudomonas putida* strain KT2440 and Gram-positive *Bacillus atrophaeus* strain DSM675 with their co-culture. Sixty-three proteins were detected as differentially expressed (at least two-fold expression change) in the co-culture relative to the pure cultures. These differentially expressed proteins were classified according to the biological processes where they are involved using Gene Ontologies. As determined by the iTRAQ quantitative labeling approach, differential expression ratios (binary vs. pure culture) ranged from -10 to 9. When comparing the mixed culture to the base cultures, more proteins were up-regulated in response to the perturbation (38 in the perturbed *B. atrophaeus* culture and 48 in the *P. putida* perturbed culture). Among the proteins with the largest expression changes in *B. atrophaeus* are elongation factor G (9-fold lower in the binary culture), acetolactate synthase III large subunit (8.5-fold lower), quinone oxidoreductase (5.8-fold lower), putative cold-shock DNA-binding domain protein (4.6-fold higher), and heat shock protein Hsp20 (4.9-fold higher). For *P. putida*, the largest expression changes were found for outer membrane lipoprotein OprI (4.5-fold lower in the binary culture), isocitrate lyase (4.3-fold lower), and DNA-binding protein HU, form N (9.7-fold higher). Approximately two-thirds of the proteins identified were involved in metabolism, and the majority of those were related to amino acid and protein metabolism, and well as nucleic acid and DNA metabolism. This is the first use of proteomics to reveal the interactions between microorganisms.

See more of this Session: Systems and Synthetic Biology of Interacting Microorganisms
See more of this Group/Topical: Food, Pharmaceutical & Bioengineering Division

(67e) Dynamic Flux Balance Modeling of a Microbial Co-Culture for Efficient Batch Fermentation of Glucose and Xylose Mixtures

Timothy Hanly[1], Morgan Urello[2] and Michael A. Henson[1], (1)Chemical Engineering, University of Massachusetts, Amherst, MA, (2)Columbia University

A requirement for the economically viable production of fuels from cellulosic biomass is the efficient consumption and conversion of its constituent sugars. Genetically engineering a single organism to metabolize multiple sugars typically results in inefficiencies due to diauxic growth or limitations in substrate uptake. Here, we present a model of a synthetic consortium for the production of ethanol composed of wild-type Saccharomyces cerevisiae and Escherichia coli strain ZSC113, two microbes that will specifically uptake glucose and xylose respectively. Dynamic flux balance analysis is employed to compare this co-culture to mono-cultures of the strains of the component microbes capable of consuming both sugars in terms of ethanol productivity. The effects of altering the amount of each microbe present in reactor inoculum and changing the time at which the batch is switched from aerobic to anaerobic cultivation are investigated. Through these process engineering strategies, a nearly two-fold increase in ethanol productivity over pure cultures in silico is achieved. Future work will focus on verifying these results experimentally to test the validity of assumptions made in our model.

See more of this Session: Systems and Synthetic Biology of Interacting Microorganisms
See more of this Group/Topical: Food, Pharmaceutical & Bioengineering Division

(67f) Tunable Synthetic Microbial Consortia: Foundations and Applications

Alissa R. Kerner, Chemical Engineering, University of Michigan, Ann Arbor, MI and Xiaoxia (Nina) Lin, Department of Chemical Engineering, University of Michigan, Ann Arbor, MI

Microbial symbiosis confers many advantages to its participants and has the potential to be utilized beyond naturally occurring communities. One application for this kind of synthetic consortium is the microbial production of molecules that are difficult to synthesize or may take many different steps and/or reactions to produce, such as biofuels or pharmaceuticals. Here we report the construction of a tunable synthetic microbial consortium created via the genetic engineering of *E. coli* auxotrophs to cross-feed and support each other when grown in co-culture. Genetic and metabolic circuits were designed to allow tunable control of the system via the export of essential amino acids by each of the auxotrophs. In this particular system, a tyrosine auxotroph (Δtyr) produces tryptophan for the tryptophan auxotroph (Δtrp) and vice versa. Inducible promoters are used to control the expression of relevant target genes, which then control the amino acid synthesis and export. It has been previously shown that the growth rate and composition of such a co-culture are determined by the export rate and cellular requirement of such essential molecules. By adding specific amounts of inducers we have been able to produce a gradient of target gene expression and a co-culture with a tunable growth rate and composition.

The relatively simple bacterial community mentioned above could be extended to a more complex inter-kingdom consortium consisting of a bacterium and a fungus, which has the potential to enable efficient and cost-effective cellulosic biofuel production. We have designed such a circuit based on quorum sensing, where the fungal member would produce bacterial signaling molecules that would allow it to control the growth of the bacterium. Through these molecules (the expression of which would be tuned exogenously) the fungus and bacterium would "communicate" and be able to grow in a coordinated and tunable manner.

See more of this Session: Systems and Synthetic Biology of Interacting Microorganisms
See more of this Group/Topical: Food, Pharmaceutical & Bioengineering Division

(67g) Yeast Artificial Cell-Cell Communication and Conway's Game-of-Life in Yeast

Jingjing Sun, Dep. of Biological Engineering, M.I.T., Cambridge, MA and Ron Weiss, Department of Biological Engineering, M.I.T., Cambridge, MA

In nature, programmed cell death is often used to remove excess cells, as is seen in embryo, finger, and brain development, sea urchin metamorphosis, and in plant immune responses. Within the context of synthetic biology, coordinating cell death and survival artificially will enhance our understanding of natural systems and our ability to engineer complex cellular patterns. In addition, the ability to precisely regulate yeast cell density is essential for numerous industrial applications, such as yeast fermentation. To explore coordination of cell death and survival in a pattern formation system, we are implementing a genetic version of Conway's Game of Life. As with the computer version of the game, cells die due to overpopulation (too many neighbors) or when they are in solitude (too few neighbors). Otherwise, cells are allowed to survive and proliferate if they have the appropriate number of neighbors. Our expectation is that 2D patterns with certain predictable properties will emerge after repeated rounds of cell death and survival. By demonstrating how changes to the underlying survival rules alter the resulting global patterns, we can improve our understanding of how complex global behavior emerges from the interaction of elements governed by simple local rules. Towards this goal, three biological devices are designed and engineered, and integrated into yeast cellular machinery, namely a quorum sensing device, a low threshold killing device, and a high threshold killing device. The quorum sensing device (QS) is designed by integrating a 'sender' part and a 'receiver' part abstracted from Arabidopsis thaliana into *Saccharomyces cerevisiae*. The low threshold killing device (LT) is based on the interaction between the apoptotic protein Bax and its counterpart Bcl-xL, that is Bax alone can kill yeast cells and Bcl-xL can rescue cell death by blocking Bax from functioning. Different from the low threshold, the high threshold killing device (HT) functions by conditionally expressing Cre recombinase to delete a critical gene in cell wall integrity, which is flanked by two LoxP sites. Based on experimental results, the low threshold killing device and the high threshold killing device have a four-fold and three-fold killing efficiency respectively. Current efforts are focused on the fine-tuning of the two killing devices to have a nice survival band-pass.

See more of this Session: Systems and Synthetic Biology of Interacting Microorganisms
See more of this Group/Topical: Food, Pharmaceutical & Bioengineering Division

(125a) Engineering T-Cell Receptor Signaling Dynamics for Cell-Based Immunotherapy

Wilson W. Wong[1], Ethan Corcoran[2], Arthur Weiss[2] and Lim Wendell[1], (1)Cellular and Molecular Pharmacology, University of California, San Francisco, San Francisco, CA, (2)Department of Immunology, University of California, San Francisco, San Francisco, CA

Adoptive transfer of T-cells expressing chimeric antigen receptors (a fusion of tumor specific single-chain antibodies and intracellular signaling domains from the T-cell receptor (TCR) signaling pathway) has shown promise as an anti-tumor therapy. However, their efficacy is limited because of limited in vivo activity, T-cell exhaustion after ex vivo expansion, low persistence inside the host, and potential safety concerns. Here we are applying synthetic biology approaches to engineer and optimize the behaviors of these tumor targeted T-cells. By engineering synthetic positive and negative feedback loops in T-cells, we show that we can precisely tune the amplitude and dynamics of the T-cell response. We have demonstrated that expression of bacterial effector proteins can be used to shut down and attenuate the response. Conversely, we have also shown that we can amplify output by expressing fusions of signaling proteins that mimic an activated signaling complex. The modularity of these circuits should allow us to construct and search for altered response profiles that improve the persistence and anti-tumor activity of engineered T-cells. Our approaches complement ongoing work on chimeric antigen receptor engineering. Furthermore, this work has yielded a new set of genetic tools to control the T-cell signaling dynamics.

See more of this Session: Synthetic Systems Biology II
See more of this Group/Topical: Topical A: Systems Biology

(125b) Directed Evolution of Quorum-Sensing Dependent Transcriptional Repressors for Programmed Intercellular Communication

Jasmine Shong, Chemical and Biological Engineering, Rensselaer Polytechnic Institute, Troy, NY and **Cynthia H. Collins**, Department of Chemical and Biological Engineering, Rensselaer Polytechnic Institute, Troy, NY

One of the key challenges in the field of synthetic biology is to engineer not only single cells but also communities of cells that exhibit coordinated, population level behaviors. Synthetic biologists have employed components from acyl-homoserine lactone (AHL)-based quorum-sensing (QS) systems in their endeavors to build muticellular systems. To date, the regulators available in the toolbox of QS regulatory parts for engineering intercellular communication have been limited to transcriptional activators, most frequently the AHL-dependent activator LuxR. Transcriptional activators function by recruiting RNA polymerase (RNAp) to a promoter sequence and thus require specific interactions with RNAp. In contrast, transcriptional repressors bind to operator regions within a promoter and inhibit the initiation of transcription. To expand this toolbox of regulatory parts, we have engineered the AHL-dependent transcriptional repressor, EsaR, for use in synthetic microbial communities. Characterization of wild-type EsaR showed that it requires micromolar concentrations of AHL to achieve complete derepression, while the LuxR will activate gene expression in the presence of nanomolar concentrations of AHL. In order to generate QS repressors with sensitivities comparable to LuxR, we built libraries of esaR mutants and identified EsaR variants with increased AHL sensitivity using an ON/OFF screening system. We first screened for an absence of gene expression to identify EsaR variants that retained the ability to repress gene expression (OFF screening). We subsequently screened for reporter gene expression in the presence of nanomolar concentrations of AHL (ON screening). We have identified a number of EsaR variants that respond to a range of AHL concentrations, from 10 nM to 1 uM. These new regulatory parts will enable future work exploring the roles of different regulatory mechanisms and network architectures for intercellular communication in both natural and engineered multicellular systems.

See more of this Session: Synthetic Systems Biology II
See more of this Group/Topical: Topical A: Systems Biology

(125c) Novel, Synthetic, Tetracycline-Inducible Transcription Regulators for E. Coli

Katherine G. Volzing, Chemical Engineering and Material Science, University of Minnesota, Minneapolis, MN and Yiannis N. Kaznessis, Chemical Engineering and Materials Science, University of Minnesota, Minneapolis, MN

We are designing, constructing and characterizing novel synthetic gene regulators. These regulators range from individual inducible fusion protein and promoter pairs to complex nonlinear regulatory networks. All of the regulators are designed to be compatible in E. coli and to control gene expression by direct transcription activation. Our investigations have complementary experimental and computer modeling components which allow us to build and characterize these complex and robust synthetic systems while reducing the required expense and effort to construct future networks. Our initial focus has been on designing, constructing and characterizing two-part synthetic regulatory systems. They are composed of an inducible fusion regulatory protein and a complementary fusion promoter. The fusion proteins are novel combinations of the inducible DNA binding tetracycline repressor (TetR) or its derivative reverse-TetR (rTetR), and the transcription activation domain LuxRΔN(2-162) (LuxRΔN). The fusion promoter contains the complementary operator sequences, tetO and luxbox, for TetR/rTetR and LuxRΔN respectively. We are experimentally characterizing the systems' gene expression regulation using real-time PCR, immunoblotting, culture fluorescence and flow cytometry while making small adjustments for optimization. In parallel, we are building and testing stochastic models with the aim of accurately capturing the systems' behavior, independent of context. Real-time PCR results indicate that the rTetR-LuxRΔN facilitates significant transcription upregulation upon induction by anhydrous tetracycline (aTc). We also observe a parallel increase in protein synthesis, monitoring mean culture fluorescence and individual cell analysis by flow cytometry. After induction with aTc, the transcription activator upregulates target gene expression, initiating the emergence of a cell population with GFP expression 2 orders of magnitude above the uninduced population. The activity of the second regulator, TetR-LuxRΔN, is inversely related to aTc levels. TetR-LuxRΔN facilitates significant GFP upregulation in the absence of aTc, while only low expression is observed in the presence of aTc. This behavior was also monitored by mean culture fluorescence and individual cell analysis by flow cytometry. In cultures with this protein, in the absence of aTc, a second cell population emerges with GFP expression increased by one order of magnitude above the untreated cultures. Next, we will investigate the behavior of each regulator in the presence of either TetR/rTetR, aiming to reduce background GFP expression for both systems. As each synthetic regulator is being characterized experimentally, we are also developing mathematical models to accurately capture the observed experimental behavior. Each system is represented by a network of biochemical reactions that describe the bimolecular interactions of transcription and translation in E. coli, while also accounting for molecule transport into the cell and molecule degradation. We are using kinetic parameters and species concentrations from the literature, and integrating these networks, their parameters, and the simulations with Synthetic Biology Software Suite (www.SynBioSS.org). The long term aim of these combined experimental and computational efforts is to build a context independent stochastic computer model for gene expression. By modeling biological systems prior to constructing them experimentally, scientists will be able to rationally design complex regulators and regulatory networks. Highly robust systems may then be synthesized cheaply and quickly to display desired functions and phenotypes.

See more of this Session: Synthetic Systems Biology II
See more of this Group/Topical: Topical A: Systems Biology

(125d) Model-Driven Minimization of the B. Subtilis Genome

Christopher Henry[1], Kosei Tanaka[2], Jenifer Zinner[1], Matthew Cohoon[1], Rick Stevens[1] and Philippe Noirot[2], (1)Mathematics and Computer Science, Argonne National Laboratory, Argonne, IL, (2) Microbial Genomics, INRA, Paris, France

Bacillus subtilis is a gram positive, sporulating bacteria often utilized in industry as a producer of high quality enzymes and proteins [1]. The flexible growth conditions, rapid growth rate, and natural competence of B. subtilis make it an ideal candidate for scientific study and industrial use. Yet many aspects of B. subtilis are still poorly understood, and the complexity of this organism makes experimental analysis and reengineering difficult. To enhance our understanding of B. subtilis and develop a more malleable and productive strain for use in a wide variety of industrial applications, we are systematically deleting all regions of the B. subtilis chromosome that are dispensable for growth on rich defined media [2]. We have applied bioinformatics techniques [3] and experimental knowledge [4] to design a set of 157 intervals of the B. subtilis genome that are predicted to be dispensable. These intervals (which range from 2 to 130 Kb in size) include all but ~832 (20%) of the genes in the B. subtilis genome. The intervals were individually deleted at 37°C in rich defined medium. 135 intervals were found to be dispensable, with the remaining 21 being essential for viability. To identify the coessential genes responsible for the 21 nonviable deletions, these intervals were iteratively subdivided into 140 smaller intervals, with new coessential and essential genes in B. subtilis ultimately being identified. All 297 experimentally implemented deletions were simulated in silico using the iBsu1103 metabolic model of B. subtilis [5]. Initially, the model incorrectly predicted the outcome of 60 (20%) interval deletions. The model was subsequently adjusted using a combination of computational analysis and experimental study to correct nearly all erroneous predictions. In several cases, the model assisted in the interval splitting process by predicting coessential gene sets. The model also predicted additional nutrients that could be added to the rich defined media to restore viability for some unviable intervals. Phenotypes for the 135 successfully implemented deletions were also studied experimentally and computationally predicting and measuring growth in minimal media, rich defined media, and LB media. Insights gained from the initial 297 interval deletions and the subsequent combination of interval deletions into a 1.5 MB knockout mutant will be discussed.

1.) Zweers JC, Barak I, Becher D et al (2008) Towards the development of Bacillus subtilis as a cell factory for membrane proteins and protein complexes. Microb Cell Fact 7:10

2.) Fabret C, Ehrlich SD, and Noirot P (2002) A new mutation delivery system for genome-scale approaches in Bacillus subtilis. Mol Microbiol 46(1):25-36

3.) Overbeek R, Disz T, and Stevens R (2004) The SEED: A peer-to-peer environment for genome annotation. Communications of the ACM 47(11):46-51

4.) Kobayashi K, Ehrlich SD, Albertini A et al (2003) Essential Bacillus subtilis genes. Proc Natl Acad Sci U S A 100(8):4678-83

5.) Henry CS, Zinner J, Cohoon M et al (2009) iBsu1103: a new genome scale metabolic model of B. subtilis based on SEED annotations. Genome Biology 10:R69

See more of this Session: Synthetic Systems Biology II
See more of this Group/Topical: Topical A: Systems Biology

(125e) A Microrna Based Synthetic Network as Stable Expression Unit in Mammalian Cells

Leonidas Bleris[1], Zhen Xie[1], Eduardo Sontag[2] and Yaakov Benenson[1], (1)Center for Systems Biology, Harvard University, Cambridge, MA, (2)Department of Mathematics, Rutgers University, New Brunswick, NJ

The utility and scalability of complex synthetic gene networks is hampered by fluctuations in stoichiometry between different gene products in individual cells. A critical contribution to the fluctuations arises from variation in the basic transcription efficiency of a gene product that in turn is determined by the cellular environment, the DNA state, and most importantly by the gene's copy number. This is typically the case when the expression unit consists of a promoter combined with an open reading frame. Therefore it is critical to construct sophisticated expression units whose gene product will depend only weakly on the number of unit copies in a cell and on the global transcription efficiency. In other words, the output protein expression will not depend on the number of DNA molecules that code for this protein, a property we call adaptation. This would raise a number of interesting hypotheses: (a) that such circuitry might be found in nature when a constant level of gene expression is required, (b) that such circuits could become valuable tools in the construction of increasingly complex synthetic circuits because they will contribute to overall robustness of the system by decreasing natural variability in the circuits' components, and (c) that such circuits could enable finely controlled gene delivery methods in gene therapy applications. Here we show that a specific feed-forward loop network topology functions as a stable expression unit in mammalian cells. We utilize transcriptional and post-transcriptional machinery and show adaptivity of an output compared to a promoter-ORF combination. We used transient transfections to characterize the units, and developed a modeling approach that takes sources of fluctuations into account and generally reproduces raw experimental data.

See more of this Session: Synthetic Systems Biology II
See more of this Group/Topical: Topical A: Systems Biology

(125f) Systematic Reduction of Models of Mechanistic Template Polymerization Processes and Applications to Genetic Networks

Vassily Hatzimanikatis, Institute of Chemical Sciences and Engineering, Ecole Polytechnique Fédérale de Lausanne (EPFL), Lausanne, Switzerland, Luis Mier-y-Teran R, Physics and Astronomy Department, Northwestern University, Evanston, IL and Mary Silber, Engineering Sciences and Applied Mathematics Department, Northwestern University, Evanston, IL

The engineering of gene regulatory circuits necessitates the use of systematic mathematical modeling to provide guidelines and make quantitative predictions. The components of such circuits pose exceedingly diverse and rich forms of interactions, it is therefore indispensable that mathematical models incorporate the essential details of the processes involved for making accurate predictions.

It has been shown that incorporating time delays in reduced models of gene networks is often essential to capture the whole range of behavior. From a mechanistic model for protein translation in the form of a large system of ordinary differential equations (ODEs), we systematically derive a reduced time-delay model by performing systematic approximations on the ODE system. We find quantitative agreement in protein synthesis rates and ribosome density on the messages between the two models. The region in parameter space where self-sustained oscillations exist is found, via bifurcation analysis, for both the reduced model obtained and a heuristic model proposed else- where. Qualitative differences are found between the models and explained in terms of the undergoing biological process.

See more of this Session: Synthetic Systems Biology II
See more of this Group/Topical: Topical A: Systems Biology

(180a) Systems Analysis of the Fermentative Metabolism of Escherichia Coli

Ramon Gonzalez[1], Abhishek Murarka[2], James M. Clomburg[2], Sean Moran[3] and Jacqueline V. Shanks[4], (1)Departments of Chemical & Biomolecular Engineering and Bioengineering, Rice University, Houston, TX, (2)Chemical and Biomolecular Engineering, Rice University, Houston, TX, (3) Biochemistry and Cell Biology, Rice University, Houston, TX, (4)Chemical and Biological Engineering, Iowa State University, Ames, IA

The fermentative metabolism of Escherichia coli has been extensively studied and represents the primary metabolic mode of interest for industrial applications. However, most of this knowledge originates from the study of metabolic components in isolation and is rather incomplete. In contrast to previous studies, we have used a system-level approach that provides an understanding of fermentative metabolism otherwise not achievable by the use of classical biochemical and genetic approaches. To this end, we used in silico and in vivo metabolic flux analysis (MFA) to investigate the fermentative utilization of glucuronate, glucose, and glycerol. These three substrates were chosen because of their differences in the oxidation state of carbon, which in turn largely dictates the distribution of metabolic fluxes in central metabolism and fermentative and biosynthetic pathways.

The fermentative metabolism of glucuronate and glucose was investigated with emphasis on the dissimilation of pyruvate via pyruvate formate-lyase (PFL) and pyruvate dehydrogenase (PDH). Prior to our studies the in vivo activity of PDH in anaerobic cultures was considered negligible and the role of this enzyme in the fermentative metabolism of E. coli remained unknown. However, we found that the growth of a PDH-deficient strain on either glucose or glucuronate was significantly impaired. Metabolite balancing, biosynthetic 13C labeling of proteinogenic amino acids, and isotopomer balancing all indicated a large increase in the flux of the oxidative branch of the pentose phosphate pathway (ox-PPP) in response to the PDH deficiency. Since both the ox-PPP and PDH generate CO_2 and reducing equivalents, the aforementioned findings led the hypothesis that the role of PDH is to provide CO_2 (metabolism of glucose) and biosynthetic power (metabolism of glucuronate) for cell growth. Biochemical and molecular genetic approaches confirmed the proposed roles of PDH. In vivo and in vitro MFA were also used to investigate the fermentative metabolism of the highly reduced substrate glycerol. This analysis confirmed a previously reported metabolic model in which the synthesis of both ethanol and 1,2-propanediol are required for the anaerobic fermentation of glycerol. The system level analysis predicted an important role for pathways not previously reported to be involved in the fermentative utilization of glycerol. These predictions were validated experimentally, thus further underscoring the importance of using a systems level approach to study microbial metabolism.

Murarka, A., Clomburg, J., and Gonzalez, R. (2010). Metabolic flux analysis of wild-type Escherichia coli and mutants deficient in pyruvate-dissimilating enzymes during fermentative metabolism of glucuronate. Microbiology-SGM 156 (6) (doi 10.1099/mic.0.036251-0).

Murarka, A., Clomburg, J. M., Moran, S., Shanks, J.V., and Gonzalez, R. (2010). Metabolic analysis of wild-type Escherichia coli and a pyruvate dehydrogenase (PDH)-deficient derivative reveals the role of PDH in the fermentative metabolism of glucose. (MS in Review)

See more of this Session: Systems Biotechnology
See more of this Group/Topical: Food, Pharmaceutical & Bioengineering Division

(180b) Gravitational Effects On Biofilm Formation by *Pseudomonas Aeruginosa*

Woo Seong Kim[1], Farah Tengra[1], Jonathan Dordick[2], Joel L. Plawsky[2] and **Cynthia H. Collins**[1], (1) Department of Chemical and Biological Engineering, Rennselaer Polytechnic Institute, Troy, NY, (2) Department of Chemical and Biological Engineering, Rensselaer Polytechnic Institute, Troy, NY

There is an urgent need to understand the effects of microgravity on the growth, cellular physiology, and cell-cell interactions in microbial biofilms. This information can then be used to curtail harmful activities of microbial consortia thriving as biofilms on the International Space Station and is essential for the long-term success of human space exploration. Bacterial biofilms were abundant on the MIR space station and were responsible for increasing corrosion and blocking a water purification system. Health and safety hazards linked to the development of biofilms are also of great concern because of the decreased in immune function observed in space travelers. We are using microarrays, laser scanning confocal microscopy, image analysis and the measurement of viable cell counts to develop a systems-level model of how gravity alters biofilm formation with the goal of developing new strategies to reduce their impact on the operation of spacecrafts and the health of their crew. We are focusing on *Pseudomonas aeruginosa* because it forms biofilms both inside and outside of the human body, can switch between benign and pathogenic interactions with humans, and may be relevant to crew health during extended missions. We are studying how gene expression and the three-dimensional structures of biofilms formed by these microbes differ in microgravity compared to normal gravity using both simulated microgravity and spaceflight experiments. We will present our recent results comparing biofilm formation on membranes in normal gravity, simulated microgravity and in spaceflight experiments.

See more of this Session: Systems Biotechnology
See more of this Group/Topical: Food, Pharmaceutical & Bioengineering Division

(180c) Combined Experimental and Computational Approach to Elucidating the Effects of Cellular Proliferation On Gene Transfer Efficiency

Charles Xue, Biomedical Engineering, University of California, Irvine, Irvine, CA and **Young Jik Kwon**, Pharmaceutical Sciences, Chemical Engineering and Materials Science, Biomedical Engi, University of California, Irvine, Irvine, CA

Cellular proliferation plays determining roles in gene delivery and it has been speculated that transfecting (or transducing) actively dividing cells by most gene vectorswould be much more efficient than quiescent cells. For example, retroviral vectors can transduce only actively dividing (mitotic) cells, which makes them a poor gene carrier in vivo where cells do not rapidly proliferate. Quantitative gene transfer efficiency of both viral and nonviral vectors to the cells under heterogeneous proliferation conditions (like in the body) has not been fully elucidated. In addition, usually gene transfer efficiency cannot be quantified until cells express transgene, requiring post-transfection (or transduction) incubation; therefore, overall gene delivery efficiency at a certain post-transfection (or transduction) time could be highly over- or under-estimated. Quantitatively understanding the above-mentioned parameters cannot be achieved by experiments alone. A431 human epithelial carcinoma and T98G human glioblastoma cells were cultured to form colonies where different levels of contact inhibition forces are applied, depending on locations. The cells toward the center of the colony no longer are able to divide in the confined area, while the cells in periphery actively spread and proliferate. A computer-simulated stochastic cell growth model was utilized to fit experimental cell growth kinetics. By employing gene expression rates collected at various post-tranfection (or transduction) time points as well as the simulated model, the specific gene transfer rate to actively dividing and non-dividing cells at the time of transfection (or transduction) were also determined. The results demonstrated that contact inhibition did have a dramatic effect on the level of gene transfer efficiency of retroviral, lentiviral, adenoviral, adeno-associated viral, and nonviral vectors; however, this effect varied significantly among the various vectors. For example, surprisingly, adenovirus transduced the non-dividing cells more efficiently than dividing ones. The results of this study indicate that a stochastic cell growth model in combination with experimentally determined gene transfer efficiencies is a useful tool in obtaining insightful findings that could provide new paradigms in developing efficient vectors for gene therapy.

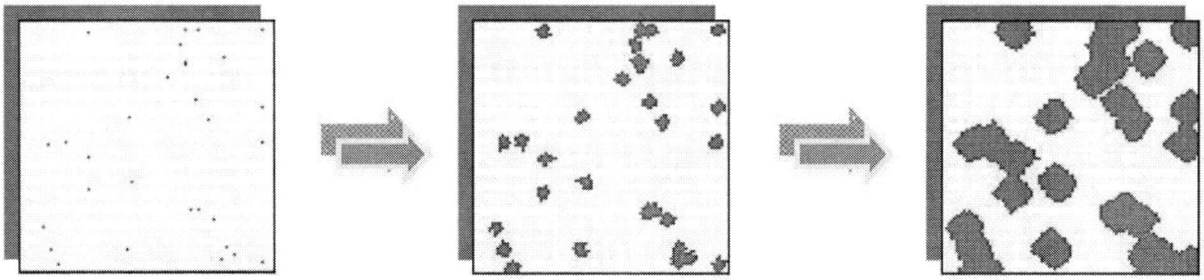

Figure. Image output from the simulation representing the progression of cell division over time. Red areas indicate contact inhibited cells and blue areas are actively dividing cells.

(180d) Role of Redox Balancing and Reactive Oxygen Species in the Metabolic Phenotype of Tumor Cells

Taylor A. Murphy and Jamey D. Young, Chemical and Biomolecular Engineering, Vanderbilt University, Nashville, TN

Reprogramming of glucose and glutamine metabolism is a key event in malignant transformation. However, the functional significance and molecular mechanisms underlying these adaptations are only now starting to emerge. Tumor metabolism is considered "wasteful" from the carbon and energetic standpoint because it produces large amounts of lactate, even when oxygen is readily available to support respiration. However, an often overlooked role of central metabolism is to supply reductant in the form of NADPH, which is typically the limiting resource in anabolic processes such as lipid and nucleotide biosynthesis. Furthermore, rapidly proliferating cells must tightly regulate reactive oxygen species (ROS) accumulation in order to avoid DNA damage while maintaining activation of MAPK signaling cascades. These critical metabolic functions require continuous production of cytosolic NADPH to effectively modulate ROS levels.

We are applying systems approaches to elucidate the role of cell-wide metabolic reprogramming to maintain redox homeostasis while optimizing intracellular ROS levels in tumor cells. We will present results from 13C metabolic flux analysis (MFA) applied to both normal FL5.12 B-cells and P493-6 lymphoma cells. These cell lines have been genetically modified to overexpress key oncoproteins including Akt or Myc as well as important metabolic enzymes such as Glut1 or hexokinase. This approach allows us to quantify how specific oncogenic signals redirect metabolic fluxes through NADPH- and ROS-producing pathways. We have observed that increased expression of the Myc oncoprotein results in a dose-dependent increase in ROS accumulation in P493-6 cells, which is associated with increased growth and upregulation of both glucose and glutamine metabolism. We will present results from 13C MFA studies aimed at elucidating the global metabolic alterations underlying these phenotypes, as well as how they respond to targeted interventions.

See more of this Session: Systems Biotechnology
See more of this Group/Topical: Food, Pharmaceutical & Bioengineering Division

(180e) Deciphering Translation - a Genome-Wide Analysis of Translation Limitations and Competitions in Saccharomyces Cerevisiae

Julien Racle[1], Amit Mehra[2] and **Vassily Hatzimanikatis**[1], (1)Laboratory of Computational Systems Biotechnology, Ecole Polytechnique Fédérale de Lausanne (EPFL), Lausanne, Switzerland, (2)Vendavo, Inc., Chicago, IL

Protein synthesis is one of the central elements in every living cell. For this process, the mRNAs coding for different genes are simultaneously competing for the same translation machinery, i.e. for the same ribosomes and for the same protein building blocks, the amino acids. But the ultimate determinants of cellular functions are the proteins, it is therefore of crucial importance to understand well this translation process. Nowadays, it is possible to measure the changes in the levels of all mRNAs and proteins in the cell, under different environmental and genetic perturbations. The large amount of data from transcriptomic and proteomic studies requires development of efficient systems-wide mathematical and computational frameworks to analyze and interpret the data.

For this purpose, we developed a genome-wide, mechanistic, mathematical model for the translation machinery in *S. cerevisiae*. This model includes information on the translation state of each mRNA (i.e. polysome size of each mRNA), and mRNA length, and can be used to estimate genome-scale kinetics of the protein synthesis machinery. It is then used to quantify the rate limiting steps of the translation process and gain insights into the mechanism of regulation of expression between different proteins. This framework allows us therefore to quantify the effect of a change in the total number of ribosomes, or also the effect of varying the level of a single mRNA on the various proteins. This information is then used to determine functional couplings in the yeast genome. We find that while 95% of the mRNA species are initiation and elongation limited, approximately 5% are elongation limited and have higher specific protein synthesis rates. We also observed that the mRNAs with ribosomal coverage greater than 0.8 display a decrease in the specific protein production rate with increasing coverage, and can be acting as sinks for system resources like ribosomes. It was also possible to determine which mRNA species have the most / least influence on the other, and we found that the most influential mRNA species are linked to proteins which form structural constituents of ribosomes, proteins related to translation elongation factor activity or key enzymes of the central carbon metabolism. Thus a perturbation to these key mRNA species has the maximum impact on protein synthesis from other mRNA species.

Our results thus have implications in the understanding of the translation machinery response to perturbations and of the global systems response.

See more of this Session: Systems Biotechnology
See more of this Group/Topical: Food, Pharmaceutical & Bioengineering Division

(180f) DILI-Sim: A Mechanistically Based Computational Model for Predicting Drug Induced Liver Injury (DILI)

Brett A. Howell[1], Melvin E. Andersen[2], Sudin Bhattacharya[2], Harvey J. Clewell III[3], Alison Harrill[1], Catherine Lisa Kurtz[1], Paul B. Watkins[1], Yuching Yang[2], Richard Ho[4], Rukmini Kumar[4], Scott Q. Siler[4] and Leif Wennerberg[4], (1)The Hamner - UNC Institute for Drug Safety Sciences, Research Triangle Park, NC, (2)Division of Computational Biology, The Hamner Institutes for Health Sciences, Research Triangle Park, NC, (3)Center for Human Health Assessment, The Hamner Institutes for Health Sciences, Research Triangle Park, NC, (4)Entelos, Inc., Foster City, CA

Drug-induced liver injury (DILI) is the adverse drug event that most frequently leads to termination of clinical development programs, as well as to regulatory actions on drugs, including failure to gain approval for marketing, restrictions in labeling, and withdrawal from the market. Of particular concern is the rare or "idiosyncratic" DILI that may only become evident late in clinical development or after a drug is marketed. This concern is driving increases in the size and duration of many Phase 3 clinical trials and creating a bottleneck in the delivery of new medicines to patients.

DILI-sim is a predictive model based on physiological processes involved in the initiation and propagation of DILI. The Hamner-UNC Institute for Drug Safety Sciences and Entelos, Inc., have partnered to develop DILI-sim in the PhysioLab® biosimulation platform. The Food and Drug Administration (FDA) has endorsed the effort and plans to use DILI-sim as a tool for recognizing drugs with a high propensity for DILI. Some of the key aspects of DILI to be addressed with DILI-sim include species differences in susceptibility to DILI, pharmacologic class versus structure-specific DILI, distinction between benign liver chemistry changes and those that indicate serious DILI potential, genetic and environmental factors (e.g. diet and disease) that may increase DILI susceptibility, and biomarkers that may identify subsets of patients who are susceptible to DILI. Acetaminophen (APAP), isoniazid, and valproic acid were chosen as exemplar drugs around which to construct a multiple pathways model. Multiple species, including but not limited to mice, rats, and humans are included. In vivo and in vitro data from rodents and humans at therapeutic doses and doses known to cause DILI were used for parameter optimization and model validation. Following development, DILI-sim will be tailored to predict idiosyncratic DILI during the clinical trial phase and for novel compounds during the early development phase. The characterization of APAP induced DILI was the primary task during phase I of development.

A physiologically based pharmacokinetic (PBPK) model was developed to describe APAP absorption, distribution, metabolism, and excretion. The metabolic pathways of glucuronidation, sulfation, and CYP oxidation to the reactive metabolite N-acetyl-p-benzo-quinone imine (NAPQI) were included. Multiple dosing routes, including intravenous, intraperitoneal, and oral were also incorporated. The PBPK model was linked to a glutathione depletion model to account for oxidative stress from NAPQI. The cell life cycle was modeled, including the ability of cells to recover from stress or suffer apoptosis or necrosis. Cell regeneration rates have been proposed as key factors in liver recovery from DILI versus acute liver failure, and were therefore also considered. Biomarkers such as alanine aminotransferase (ALT), aspartate aminotransferase (AST), alkaline phosphatase (AP; ALP), gamma-glutamyl transferase (GGT), albumin, prothrombin time, lactate dehydrogenase (LDH), bilirubin, (direct/conjugated and total), as well as several novel markers, are represented as model outputs.

The model accurately reproduced APAP pharmacokinetic measures for rats, mice, and humans. Several key findings concerning APAP induced DILI have already been uncovered. The depletion of 3' -phosphoadenosine 5'phosphosulfate (PAPS) is crucial to accurately reproducing the sulfation pathway in rats. Metabolism differences across species explain much of the variation in species specific susceptibility to APAP induced DILI. Several hypotheses put forth in literature concerning the innate immune response during DILI were evaluated for plausibility. One major area addressed is the significant difference between APAP metabolism during therapeutic doses, which most APAP modeling efforts focus on with respect to humans, and clinical overdose scenarios, which are far more important for DILI. Finally, a variety of subject-specific attributes were varied in search of factors increasing a patient's risk for developing APAP induced DILI.

See more of this Session: Systems Biotechnology
See more of this Group/Topical: Food, Pharmaceutical & Bioengineering Division

(180g) Micro-Droplet Enabled Co-Cultivation of Symbiotic Bacteria

Jihyang Park[1], Mark A. Burns[1] and Xiaoxia (Nina) Lin[2], (1)Chemical Engineering, University of Michigan, Ann Arbor, MI, (2)Department of Chemical Engineering, University of Michigan, Ann Arbor, MI

In nature, most microbes live in synergistic communities as a way to adapt to and thrive in their environments, such as ocean, soil, and higher organisms as hosts. These microbial communities play important roles in a wide spectrum of ecosystems and form diverse interactions with one another and with their surroundings. Microbial interactions in natural microbiota are, in many cases, crucial for the sustenance of the communities, but these interactions remain largely unknown because of the inherent complexity and difficulties in laboratory cultivation. Most of previous works were based on genetic identification, while laboratory co-cultivation for elucidating microbial intercellular networks have been hardly investigated so far. In this work, we developed a simple microfluidic device for highly parallel co-cultivation of symbiotic microbial communities and demonstrated its effectiveness in discovering synergistic interactions among microbes. Using aqueous micro-droplets, which were serially generated in Parylene-coated glass devices adapting slanted T-junction geometry and dispersed in a continuous oil phase, the device could readily encapsulate and co-cultivate various subsets of a microbial community. A large number of droplets, up to ~1400 in a 10μ mx5μ m chamber, were generated within 3 seconds and packed very tightly in the chamber. A synthetic model system consisting of cross-feeding E. coli mutants was used to mimic various compositions of symbionts and other microbes in natural microbial communities. Our device was able to detect pair-wise symbiotic relationship when one partner accounted for as low as 1% of the total population or each symbiont was about 3% of the artificial community. In addition, murine fecal microbial samples were cultivated on the device and compositions of grown cultures were analyzed. Off-chip PCR of 16S rRNA genes with universal bacterial primers was carried out on retrieved co-cultures to determine species profiles of the cultivated subsets of the microbiota. It was shown that only a small number of species from the original sample were able to grow together in compartmentalized droplets under the tested conditions. This parallelized microfluidic compartmentalization and co-cultivation platform can also be extended for the investigation of microbial communities in other environments.

See more of this Session: Systems Biotechnology
See more of this Group/Topical: Food, Pharmaceutical & Bioengineering Division

(233h) Cytomic and Lipidomic Analysis of Snow Algae Chlamydomonas Nivalis Under NaCl Stress by Laser Scanning Confocal Microscope, Flow Cytometry and GC-MS

Dong Wei, Xiao-li Jiang and Na Lv, School of Light Industry and Food Sciences, South China University of Technology, Guangzhou, China

Snow algae Chlamydomonas nivalis, a typical microalgal species discovered on snow field at Antarctic and similar extreme environments on earth, can adapt to very low temperature, high light intensity and strong UV radiation for survival by forming aplanospore rich in astaxanthin, a brown red pigment with super anti-oxidative effect. C. nivalis is realized not only a model species to investigate the mechanism of cell response and adaptation to stress conditions, but also a potential alternative bio-resource for astaxanthin production by microalgal biotechnology. The response and resistance of microalgae to stress environment depend on a series of essential regulation of metabolic pathway, especially on lipid biosynthesis and cell wall structure. However, current research of C. nivalis mostly focuses on the classification, life cycle, modes of reproduction, regional distribution and cell ultra-structure. Lipid profile of C. nivalis in resistant physiology has paid little attention.

In the present work, the 6-day culture of C. nivalis UTEX LB 2824 in Bold 1NV medium under autotrophic growth was treated by NaCl with final concentrations of 0.25%, 0.50%, 0.75%, 1.00%, 1.25%, 1.50% for 1, 2, 3, 5, 7, 11, 15, 24 or 48 h, respectively. The content of neutral and polar lipid referred as fluorescent intensity in stressed cells after stained with specific dye Nile Red was quantified by cytomic analysis based on laser scanning confocal microscope (LSCM, Ex: 543 nm; scan: 560-615 nm) and flow cytometry (FCM, Ex: 488 nm, FL1 at 530 nm, FL2 at 575 nm and SSC signal). After extraction and derivation of triacylglycerol (TG) and free fatty acids (FFA) from stressed cells, the profiles of FFA and FA in TG were analyzed by GC-MS. In general, the content of neutral and polar lipid showed the similar trends with total lipid in cells. The fluorescent intensities of lipids constantly increased to the maximum during 7 h, then dropped to a stable level within 24 to 48 h. The trend was not rigorous within all NaCl concentrations, but the highest increase of 211% for total lipid at 0.75% NaCl for 7 h, 68 times increase for neutral lipid at 1.00% NaCl for 7 h and 10 times increase for polar lipid at 1.25% NaCl for 5 h were observed, respectively. The cell size in all groups of NaCl stress enlarged quickly during 3 to 5 h, then stopped to become large. Big changes of composition and content of several FFA (C12:0, C16:0, C16:3, C18:0, C18:1 and C18:3) were observed with variation for 4.61, 3.70, 4.37, 2.68, 2.64, 3.14 times, respectively. C16:0, C18:0, C18:1 and C18:3 in TG also varied much for 2.67, 2.53, 4.91, 2.42 times, respectively. The results indicated C. nivalis could synthesize and accumulate lipid and change lipid profiles very quickly during several hours by stress induction. The content of C16:0, C18:0, C18:1 and C18:3 could change the degree of lipid unsaturation (DLU) and play an important role on regulating the membrane fluidity, which result an enhancement of cell response and resistance to NaCl stress. This is the first report to demonstrate the lipid profile changes corresponding to NaCl stress by "omics" technology, and figure out the regulation mechanism of lipid biosynthesis by NaCl induction in C. nivalis.

See more of this Session: Multiscale Systems Biology
See more of this Group/Topical: Topical A: Systems Biology

(233b) Experimental and Mathematical Approaches to Multi-Scale Analysis of Tumor Necrosis Factor-Regulated Immune Response to Tuberculosis

Mohammad Fallahi-Sichani[1], Mohammed El-Kebir[2], Simeone Marino[3], Denise E. Kirschner[3], Steven L. Kunkel[4] and Jennifer J. Linderman[1], (1)Chemical Engineering, University of Michigan, Ann Arbor, MI, (2)Center for Integrative Bioinformatics, VU University Amsterdam, Amsterdam, Netherlands, (3) Microbiology and Immunology, University of Michigan Medical School, Ann Arbor, MI, (4)Pathology, University of Michigan Medical School, Ann Arbor, MI

Tuberculosis (TB) is a common and deadly infectious disease caused by a highly successful bacterium, Mycobacterium tuberculosis (Mtb). Multiple host immune factors control the formation of a self-organizing aggregate of immune cells termed a granuloma in the lungs after inhalation of Mtb. One such factor, tumor necrosis factor-α (TNF), is a protein that regulates inflammatory immune responses. Availability of TNF within a TB granuloma has been proposed to have a critical role in protective immunity against TB. However, mechanisms at different time/length scales (e.g. molecular, cellular and tissue scales) that control TNF availability in a developing TB granuloma and their significance in determining the disease outcome (primary disease, latency, or TB reactivation) are poorly understood. To study the role of TNF and TNF receptors in the immune response to Mtb, we use an artificial experimental mouse model of TB. Granuloma formation in this model is induced in mice following injection of Sepharose beads covalently coupled to Mycobacterium purified protein derivative (PPD) antigen. PPD bead granulomas form at day 2 and reach their maximal size at day 4 following intravenous injection of PPD-coated beads. We investigate the role of TNF and the two known types of TNF receptors on the immune response to Mtb by performing experiments on normal mice as well as TNF or TNF receptor knockout mice and studying the size, morphology and cellular composition of granulomas measured using flow cytometry. Further, the mouse model is used to study the effects of mouse analogues of TNF-neutralizing drugs (that are used to treat inflammatory diseases and may cause TB reactivation) on size, morphology and cellular composition of granulomas to help explain why different drugs have been identified to induce distinct effects on TB reactivation. We also compare our experimental results with our first-generation multi-scale agent-based model that describes TNF-regulated immune response to TB. This model describes TNF-regulated immune responses to Mtb in the lung and includes a variety of mechanisms at both molecular and cellular scales. Taken together, our modeling and experimental data can help to elaborate relevant features of the immune response to Mtb infection, identifying new strategies for therapy/prevention.

See more of this Session: Multiscale Systems Biology
See more of this Group/Topical: Topical A: Systems Biology

(233c) Multi-Scale Agent-Based Modeling of Human Endotoxemia

Tung T. Nguyen[1], Steven Calvano[2], Stephen Lowry[2] and Ioannis P. Androulakis[3], (1)BioMaPS Institute for Quantitative Biology, Rutgers University, Piscataway, NJ, (2)Department of Surgery, UMDNJ-Robert Wood Johnson Medical School, New Brunswick, NJ, (3)Biomedical Engineering, Chemical & Biochemical Engineering, Rutgers University, Piscataway, NJ

The systemic inflammatory response syndrome (SIRS) often accompanies critical illness but is evoked by many stimuli e.g. infection, trauma, invasive surgery and biological stressors in general [1]. It often progresses to severe cases of sepsis with a substantial morbidity and mortality where bacterial growth is found and organ failure occurs [2, 3]. Despite increasing knowledge about pathophysiological pathways and processes involved in sepsis as well as promising results on animal studies and preclinical trials [4, 5], the molecular mechanisms and physiological significance of the systemic inflammatory response are still not fully understood. In order to learn more about the mechanisms associated with the host inflammatory responses, the human endotoxemia model, an in vivo model of systemic inflammation in which a single intravenous bolus of E.coli endotoxin (LPS) is given to healthy human subjects, has been employed. The model results in many similar physiological host responses that characterize Gram-negative infection [6], providing an invaluable source for the systemic identification of biological features representing the complex dynamics of a host undergoing inflammatory responses [7, 8].

With the assumption that transcriptional signatures can be able to characterize the cellular dynamics in response to the inflammatory agent, high-dimensional transcriptional human endotoxemia data were decomposed into significantly co-expressed clusters [9], three out of which are selected as representatives for common expression profiles of inflammatory biomarkers including pro-inflammatory, anti-inflammatory, and bio-energetic transcriptional patterns. Additionally, the activation for 'translating' from extra-cellular signals to intra-cellular responses is assumed to be governed by signaling cascades activated with transcription factors (e.g. NFkB [10, 11]) that drive the expression of relevant genes. In an attempt towards a better understanding of the molecular mechanisms as well as the complex dynamics of the host inflammatory responses, we previously developed a semi-mechanistic indirect response model using the ordinary differential equation (ODE) system [12]. However, it is fully deterministic with respect to their behavior given a certain set of initial conditions and assumes the homogeneity and perfect mixing within compartments as well as ignores the spatial aspect [13, 14]. Thus, we here explore an alternative approach – agent-based model (ABM) which can naturally account for more biological phenomena e.g. stochasticity [15, 16], heterogeneity [17, 18].

ABM is an object-oriented, rule-based, discrete, and stochastic modeling method [19, 20]. Interactions between agents are nonlinear, stochastic, spatial, and are described with multiple compartments and asynchronous movements. However, since the model is relied on an ill-defined principle of 'emergence', the ABM is not a 'mathematical model' per se, and thus is not subjected to formal analysis and 'solved' [20]. One of the most important issues is the estimation of model parameters. With the assumption that the default system production of a component type is approximately equal to its degradation in the normal (healthy/steady) state, we present a novel technique based on the pharmacokinetic principles for tuning parameters including the population sizes and the synthesis rates of component types. As such, in a healthy system the total number of individuals of each component type will not change significantly over the time. After all parameters are identified to maintain the system in the healthy state, circadian rhythms are incorporated and then LPS is putatively injected with various doses for qualitatively examining the profiles of representative responses.

In conclusion, the aim of this study is two folds: (i) develop a 'better' in silico agent-based model that can provide more insights into the human endotoxemia, and (ii) utilize the created model to study the individual cellular dynamics and especially ensemble patterns that lead to distinct outcomes under the administration of different LPS doses.

References

1. Nystrom PO: The systemic inflammatory response syndrome: definitions and aetiology. J Antimicrob Chemother 1998, 41 Suppl A:1-7.

2. Annane D, Bellissant E, Cavaillon JM: Septic shock. Lancet 2005, 365(9453):63-78.

3. Angus DC, Linde-Zwirble WT, Lidicker J, Clermont G, Carcillo J, Pinsky MR: Epidemiology of severe sepsis in the United States: analysis of incidence, outcome, and associated costs of care. Crit Care Med 2001, 29(7):1303-1310.

4. Riedemann NC, Guo RF, Ward PA: Novel strategies for the treatment of sepsis. Nat Med 2003, 9 (5):517-524.

5. Deans KJ, Haley M, Natanson C, Eichacker PQ, Minneci PC: Novel therapies for sepsis: a review. J Trauma 2005, 58(4):867-874.

6. Santos AA, Wilmore DW: The systemic inflammatory response: perspective of human endotoxemia. Shock 1996, 6 Suppl 1:S50-56.

7. Calvano SE, Xiao W, Richards DR, Felciano RM, Baker HV, Cho RJ, Chen RO, Brownstein BH, Cobb JP, Tschoeke SK et al: A network-based analysis of systemic inflammation in humans. Nature 2005, 437(7061):1032-1037.

8. Talwar S, Munson PJ, Barb J, Fiuza C, Cintron AP, Logun C, Tropea M, Khan S, Reda D, Shelhamer JH et al: Gene expression profiles of peripheral blood leukocytes after endotoxin challenge in humans. Physiol Genomics 2006, 25(2):203-215.

9. Nguyen TT, Nowakowski RS, Androulakis IP: Unsupervised selection of highly coexpressed and noncoexpressed genes using a consensus clustering approach. Omics 2009, 13(3):219-237.

10. Courtois G: The NF-kappaB signaling pathway in human genetic diseases. Cell Mol Life Sci 2005, 62(15):1682-1691.

11. Ihekwaba AE, Broomhead DS, Grimley RL, Benson N, Kell DB: Sensitivity analysis of parameters controlling oscillatory signalling in the NF-kappaB pathway: the roles of IKK and IkappaBalpha. Syst Biol (Stevenage) 2004, 1(1):93-103.

12. Foteinou PT, Calvano SE, Lowry SF, Androulakis IP: In silico simulation of corticosteroids effect on an NFkB- dependent physicochemical model of systemic inflammation. PLoS One 2009, 4(3):e4706.

13. Vodovotz Y, Constantine G, Rubin J, Csete M, Voit EO, An G: Mechanistic simulations of inflammation: current state and future prospects. Math Biosci 2009, 217(1):1-10.

14. Wakeland WW, Gallaher EJ, Macovsky LM, Aktipis CA: A comparison of system dynamics and agent-based simulation applied to the study of cellular receptor dynamics. The 37th Hawaii International Conference on System Sciences 2004.

15. Raj A, van Oudenaarden A: Nature, nurture, or chance: stochastic gene expression and its consequences. Cell 2008, 135(2):216-226.

16. Raser JM, O'Shea EK: Noise in gene expression: origins, consequences, and control. Science 2005, 309(5743):2010-2013.

17. Geva-Zatorsky N, Rosenfeld N, Itzkovitz S, Milo R, Sigal A, Dekel E, Yarnitzky T, Liron Y, Polak P, Lahav G et al: Oscillations and variability in the p53 system. Mol Syst Biol 2006, 2:2006 0033.

18. Nelson DE, Ihekwaba AE, Elliott M, Johnson JR, Gibney CA, Foreman BE, Nelson G, See V, Horton CA, Spiller DG et al: Oscillations in NF-kappaB signaling control the dynamics of gene expression. Science 2004, 306(5696):704-708.

19. Chavali AK, Gianchandani EP, Tung KS, Lawrence MB, Peirce SM, Papin JA: Characterizing emergent properties of immunological systems with multi-cellular rule-based computational modeling. Trends Immunol 2008, 29(12):589-599.

20. An G: Introduction of an agent-based multi-scale modular architecture for dynamic knowledge representation of acute inflammation. Theor Biol Med Model 2008, 5:11.

See more of this Session: Multiscale Systems Biology

(233d) Multiscale Modeling and Analysis of TRPV1 Regulation in Sensory Neurons

Jeffrey D. Varner and Sang Ok Song, Chemical and Biomolecular Engineering, Cornell University, Ithaca, NY

TRPV1 (transient receptor potential vanilloid subtype 1) receptors play a central role in nociception and inflammatory pain. Interestingly, both agonists and antagonists of TRPV1 have been tested therapeutically, suggesting that the regulation of TRPV1 activity is a complex balance between pro- and anti-pain signaling states. Understanding the complex regulation of these receptors, including their integration with other cellular signaling networks, requires mathematical modeling. In this study, we developed a mechanistic mathematical model of the regulation of the different subunits of TRPV1 channels in sensory neurons. Using this model, we investigated mechanisms by which inflammatory mediators in damaged tissues might sensitize TRPV1 to stimuli. We also investigated acute capsaicin-induced calcium-dependent desensitization of TRPV1 channels. Building upon previous literature from our laboratory and others, we postulated a four-state allosteric TRPV1 channel model. We assumed the activation and deactivation states of the channel were local minima in a complex energy landscape. Ion/ Ligand binding, phosphorylation and dephosphorylation events along with external inputs such as temperature or applied voltage modulated the activation energy associated with the transition between discrete channel states. We then embedded this channel model into a larger continuum model of the regulation of TRPV1 channels by intracellular ion binding and phosphorylation/dephosphorylation. We extended our previous work on ATP- induced Ca2+ dynamics in peripheral nerve endings to include the TRPV1 model along with the ATP-induced PKC pathway, PGE2-induced PKC/PKA-dependent signaling, and the Ca2+-dependent CaMKII and calcineurin pathways. The aggregate model recapitulated and ultimately predicted features of TRPV1 channels including their combined responses to temperature, voltage and capsaicin binding. Sensitivity analysis of the aggregate model suggested that cellular signaling mechanism were more important than TRPV1 channel kinetics where Ca2+ levels were the most critical influence on TRPV1 modulation. Taken together, although limited to small part of the overall architecture of TRPV1-related pain signaling, the mathematical model presented here could be a useful starting point for further investigation of the many factors controlling TRPV1 activity and pain.

See more of this Session: Multiscale Systems Biology
See more of this Group/Topical: Topical A: Systems Biology

(233e) Evaluation of Multimodal Surface Interactions of Acidogenic Bacterial Biofilms

Michael S. Waters, Nancy Lin and Sheng Lin-Gibson, Polymers Division, Biomaterials Group, National Institute of Standards and Technology, Gaithersburg, MD

Acidogenic biofilms are well-known to be efficient colonizers of surfaces which can have devastating effects from corrosion of ships hulls to bone loss, which, in either case, leads to significant structural damage of materials that are relied upon for day to day use. In this study, we take novel multi-scale multimodal approach to gain a deeper insight into acid-producing biofilm behavior, using Streptococcus mutans as a model. S. mutans is a well studied dental pathogen and the primary etiologic agent of dental caries. The capacity of S. mutans biofilms to exert pathogenic effects are due to its ability to rapidly secrete extra-polymeric substances (EPS), and lower the local and surrounding pH, which both inhibits the growth of other biofilms and etches a protective niche in the hydroxyapatite surface of a tooth. Utilizing sucrose-dependent protein pathways, glucosyltransferases produce copious amounts of adhesive polymers (alpha 1-3 and alpha 1-6 glucans), which facilitate the formation of a rigid, immobilized biofilm. Sucrose is then converted into lactic acid, which then demineralizes and erodes the tooth surface. Although these events are generally known, the inability to temporally or spatially quantify the occurrence of biofilm-mediated damage to tooth surfaces with respect to the constantly changing properties of the causative biofilm has left many questions unanswered. Using a unique multimodal, multi-scale approach, fluorescently labeled S. mutans biofilms were evaluated for their pathogenic attributes in the presence and absence of sucrose using rheometry and high-resolution interferometric optical profilometry over time in clinically mimicked conditions. The results of this study not only provide critical insight into the mechanism of dental pathogenesis for S. mutans, but also provides an advanced means of evaluating the capacity of newly engineered dental polymers to protect the tooth surface. This experimental approach is designed to be a simple and rapid system for evaluation of the effects of surface-modifying acidogenic biofilms.

See more of this Session: Multiscale Systems Biology
See more of this Group/Topical: Topical A: Systems Biology

(233f) Intrinsic and Extrinsic Regulation of Macrophage Polarization in Immunological Microenvironments

Yishan Chuang, Chemical and Biologial Engineering, Northwestern University, Evanston, IL and **Joshua N. Leonard**, Chemical and Biological Engineering & Robert H. Lurie Comprehensive Cancer Center, Northwestern University, Evanston, IL

Immunological microenvironments play central roles in nearly every chronic disease, including cancer, infections, and autoimmune disease. Surprisingly, we know little about how the cellular networks within these environments operate. An important feature of these networks is that immune cells such as macrophages and dendritic cells can adopt alternative functional states (e.g., either pro- or anti-inflammatory), which is known as "polarization", and can interconvert between these states, a capability termed functional "plasticity". Tumors and pathogens manipulate this balance to force the local immune network into an impaired state, preventing immune clearance and posing a major barrier to therapeutic intervention. Here, we investigate the mechanisms by which macrophages regulate the polarization choice, a process that remains poorly understood. The canonical view is that external cues guide polarization in an essentially deterministic fashion. However, new evidence suggests that individual macrophages respond stochastically to identical stimuli. We have also observed that delivering "incoherent" stimuli (e.g., a mixture of molecules that each individually drives polarization towards a distinct state) induces a transient diversity of cellular states that becomes less pronounced over time. We hypothesize that polarization dynamics are regulated by a balance between intrinsic mechanisms (which may generate an initial diversity of functional states) and extrinsic mechanisms (which allow neighboring cells to collectively "vote" for one functional state or another, for example by secreting auto-regulatory cytokines to a common local pool). Such coordination between responding immune cells could provide a mechanism for managing and regulating functional variability. We have developed a system for tracking polarization in individual cells within a population in real time and experimentally manipulating interactions within these cellular networks. Here we will present data that provide new insights into this medically important phenomenon.

See more of this Session: Multiscale Systems Biology
See more of this Group/Topical: Topical A: Systems Biology

(233g) Combination of Pre-Power Stroke Compliance with Pre and Post Powerstroke Release of ADP

Richard L. Long, New Mexico State University, Las Cruces, NM

Details of excitation contraction coupling in skeletal muscle action remain under scrutiny, in part because real time measurements of molecular scale action transients remain challenging. The relationship between the nano-scale molecular events and the macroscale action is, however, largely massively parallel, which makes it possible to model the link between the nano-scale and the macro-scale in a relatively simple way as a first approximation. Further, this can be combined with the excitation process simply for appropriate values of Damkohler number for calcium activation.

Another observable issue is the filament compliance which appears to occur in a significant way in the pre powerstroke portion of the protein complex cycle.

Combination of these featues enables the integrated modeling of the protein cycle which allows for scaleup from the nnano-scale molecular response to the macro-scale tissue response. This can be used to extract nano-scale parameters from macro-scale data. These are testable model results.

See more of this Session: Multiscale Systems Biology
See more of this Group/Topical: Topical A: Systems Biology

(281a) Using Inducible Gene Expression to Characterize Oncogene Cooperation in a Murine Pre-B Cell Line

Jonathan Vroom, Joshua Ferreira and Clifford Wang, Chemical Engineering, Stanford University, Stanford, CA

Cancer is caused by multiple gene mutations that each contribute to the transformation of normal cells. There is more to be learned, however, about the sequence of this transformation and how the genes involved are cooperating to produce cancerous cells. Drastic changes in the expression of a single oncogene often initiate antiproliferative signals that lead to senescence or apoptosis. Presumably, therefore, mutated genes must be expressed at tolerable levels and occur in an order that allows them to circumvent the cells natural mechanisms of tumor suppression. In this study we focused on two main goals. First, we created a system that would allow us to quantitatively determine the effect of cancer-associated genes on cell proliferation at a range of expression levels. Second, we used this system to screen pairs of genes to find novel cooperation that may help elucidate how the underlying regulatory pathways interact.

Generally, in order to determine the function of a gene, most studies highly overexpress and/or knockdown/knockout gene expression. In fact, in mammalian cells there are few tools available to inducibly express genes in a predictable manner. In this study we used a new approach wherein the proteins of interest are fused to destabilized protein domains which are inducibly stabilized by the presence of specific small molecules. Since it operates at the protein level the clonal variability is reduced and other markers/drug resistance genes can be driven off the same promoter. Next, we looked at proliferation rates as an indicator of cell response to the induced gene expression.

We looked at the interactions of pairs of genes. To identify gene candidates, we started by looking at a large mouse lymphoma tumor screen that used murine leukemia retrovirus to initiate tumorigenesis. It was assumed that pairs of genes that were frequently found in conjunction in tumors may cooperate in the transformation of normal cells. Using the technique outlined above and the two orthogonal inducible protein control systems, we expressed pairs of genes in a v-Abl induced mouse pre-B cell line.

See more of this Session: Genomic Approaches to Systems Biology
See more of this Group/Topical: Topical A: Systems Biology

(281b) Gene Expression Profiling of Short- and Long-Term Changes in Rat Liver Following Burn Injury and CLP Treatment

Qian Yang[1], Mehmet A. Orman[2], Francois Berthiaume[3], Marianthi Ierapetritou[2] and Ioannis (Yannis) P. Androulakis[4], (1)Chemical and Biochemical Engineering, Rutgers - The State University of New Jersey, Piscataway, NJ, (2)Department of Chemical and Biochemical Engineering, Rutgers University, Piscataway, NJ, (3)Department of Biomedical Engineering, Rutgers University, Piscataway, NJ, (4) Rutgers - The State University of New Jersey, Piscataway, NJ

An important player in systemic hypermetabolism is the liver, which generally controls circulating levels of metabolites. It is well known that the hepatic response to severe injury and bacterial infections are characterized by a significant up-regulation of glucose, fatty acid, and amino acid turnover in the liver [1-4]. Furthermore, up-regulation in the expression levels of genes involved in the urea cycle, gluconeogenesis, and the metabolism of several amino acids were also reported [1, 2]. Prior studies suggested that changes in stress related inflammatory pathways leading to hypermetabolic state are due to the persistent changes in gene expression in the liver [1]. In order to understand the mechanism behind the hypermetabolic response to burn injury and sepsis, a comprehensive analysis is required, such as the gene expression basis underlying the liver's short term and long term response to the insult.

The aim of this study is to gain a better understanding of gene expression changes in the rat liver during the first 10 day after the burn and cecal ligation and puncture (CLP) treatments. 20 % total body surface area third degree scald burn is applied to the animals using boiling water. On 2nd day after burn trauma, infection is induced by CLP. Control groups are treated identically except that they are given a sham-burn using 37 C water or sham-CLP the same survival surgical procedure without any puncture in cecum. Therefore, for systematic DNA microarray analysis, four different rat groups which receive different treatments: 1) sham-burn followed 2 days later by CLP, 2) burn followed by CLP, 3) sham-burn followed by sham-CLP and 4) burn followed by sham-CLP, and the dynamics of liver-specific transcriptional responses are recorded.

In order to analyze such a large amount of data, clustering and classification approaches developed by our lab recently are explored as an essential exploratory tool for discovering biological patterns and class prediction. The gene expression of above stated each experimental group is analyzed individually by the consensus clustering [5] which intends to identify a subset of more 'clusterable' set of genes that are highly coexpressed within a set of differentially expressed genes in each group.

Our results indicate significant circadian variability of liver-specific responses, as evaluated through the sham measurements, whereas comparisons among the alternative injury models will further elucidate the dynamics of the progression of the inflammatory response and help delineate the implications of compromised host to subsequent bacterial infections.

References

1. S. Banta, M. Vemula, T. Yokoyama, A. Jayaraman, F. Berthiaume, and M.L. Yarmush, Contribution of gene expression to metabolic fluxes in hypermetabolic livers induced through burn injury and cecal ligation and puncture in rats, Biotechnology and Bioengineering, 97 (2007) 118-137.

2. M. Vemula, F. Berthiaume, A. Jayaraman, and M.L. Yarmush, Expression profiling analysis of the metabolic and inflammatory changes following burn injury in rats, Physiological Genomics, 18 (2004) 87-98.

3. K. Lee, F. Berthiaume, G.N. Stephanopoulos, D.M. Yarmush, and M.L. Yarmush, Metabolic Flux Analysis of Postburn Hepatic Hypermetabolism, Metabolic Engineering, 2 (2000) 312-327.

4. K. Lee, F. Berthiaume, G.N. Stephanopoulos, and M.L. Yarmush, Profiling of dynamic changes in hypermetabolic livers, Biotechnology and Bioengineering, 83 (2003) 400-415.

5. T.T. Nguyen, R.S. Nowakowski, and I.P. Androulakis, Unsupervised Selection of Highly Coexpressed and Noncoexpressed Genes Using a Consensus Clustering Approach, OMICS A Journal of Integrative Biology, 13 (2009) 219-237.

See more of this Session: Genomic Approaches to Systems Biology
See more of this Group/Topical: Topical A: Systems Biology

(281c) Identification of Cell-Type Specific Transcripts for Informative Diagnostics

Younhee Ko, Department of Computer Science, Institute for Genomic Biology, University of Illinois at Urbana-Champaign, Urbana, IL, Leroy Hood, Institute for Systems Biology, Seattle, WA and Nathan, D. Price, Department of Chemical and Biomolecular Engineering, Institute for Genomic Biology, University of Illinois, Urbana-Champaign, Urbana, IL

The identification of cell-type specific or brain-region specific transcripts adds specificity for the diagnosis of a wide variety of brain disorders, such as traumatic brain injury or brain cancers. One goal of identifying cell-type and brain-region specific genes is that they encode for proteins that could be found in the blood – providing potentially highly specific information about the spatial location of perturbed networks from a non-invasive proxy measurement. The development of such multi-parameter blood protein markers lies at the very heart of the predictive medicine. Thus, in this study, we have proposed a novel method - called EigenBrain - to identify novel candidate cell-type specific transcripts, and validated its performance using 10-fold cross-validation. All cases studied except the coronal section of neuron cells showed more than 95% specificity and 95% accuracy.

Our EigenBrain approach transformed the original feature vector into the new lower-dimensional EigenBrain space, which greatly improved the accuracy of our algorithm for identification of cell-type specific transcripts. We applied this approach to the set of in situ hybridization (ISH) mouse brain images of 20,000 genes in the Allen Brain Atlas dataset and discovered a strong candidate set of brain-region and cell-type specific transcripts. Furthermore, we also investigated these highly expressed patterns in brain regions for each cell type - called EigenBrain. The enrichment of functional categories and relevant pathways involved in each candidate cell-type specific transcripts were investigated for further biological validation.

See more of this Session: Genomic Approaches to Systems Biology
See more of this Group/Topical: Topical A: Systems Biology

(281d) Exploring Bistable Molecular Switches and Their Functional Roles Using a Meta-Analysis That Integrates Expression Data with Regulatory Networks

Ming Wu, Computer Science and Engineering, Michigan State University, East Lansing, MI and Christina Chan, Chemical Engineering and Materials Science, Biochemistry and Molecular Biology, Michigan State University, East Lansing, MI

An important topic in systems biology is to understand the dynamic properties of the biological networks. Previous studies have provided important insights into the mechanistic principles of dynamic behaviors by applying kinetic modeling to condition-specific and very small networks. However, it remains unclear how to bridge such dynamic behaviors described by bottom-up mechanisms with top-down global functional analysis, i.e. to associate the dynamic features of single molecules with the large-scale network structure and functionality.

Here we explore potential bistable molecular switches based on gene expression profiles across many different conditions, representing various biological states. Distinct from most approaches proposed on the meta-analysis of microarray data that attempt to identify differentially expressed genes, we analyze the expression profiles in the context of multistability as a dynamic property of kinetic systems. We apply mixture Gaussian model to characterize the bimodality in gene expression profiles and perform Expectation Maximization algorithm together with information criterions to identify potential bistable states. We hypothesize that the dynamic property of the molecular components within a network is contingent on the network organization and the resulting dynamic behavior corresponds to specific biological functions, i.e. positive feedback. Our study integrates yeast microarray, transcriptional regulatory network, protein interaction network and genetic knock-out data to reveal the relationship between bistability, local network structure and functionality, and provides proof-of-concept of our proposed meta-analysis approach.

Finally we apply our meta-analysis approach on human gene expression datasets comparing multiple cancer cell types and tissues with non-cancerous diseases, to identify potential "molecular switches" that are universal across cancer types. These common modules may define the switch-like behavior and thereby determine the status of the biological systems and whether it could be perturbed into tumorigenesis.

See more of this Session: Genomic Approaches to Systems Biology
See more of this Group/Topical: Topical A: Systems Biology

(281e) The Role of Toxin/Antitoxin Systems in the General Stress Response, Biofilm Formation, and Persister Cell Formation

Thomas K. Wood[1], Xiaoxue Wang[1], Younghoon Kim[1], Qun Ma[1], Seok Hoon Hong[1], Michael J. Benedik[2], Rebecca Page[3] and Wolfgang Peti[4], (1)Artie McFerrin Department of Chemical Engineering, Texas A&M University, College Station, TX, (2)Department of Biology, Texas A&M University, College Station, TX, (3)Department of Molecular Biology, Cell Biology and Biochemistry, Brown University, Providence, RI, (4)Molecular Pharmacology, Physiology, and Biotechnology, Brown University, Providence, RI

Almost all bacteria grow on solid surfaces or at air/liquid interfaces in complex communities known as biofilms. These communities are useful for engineering applications like the remediation of toxic chemicals and are important for medical problems such as chronic infections. We investigated the genetic basis of biofilm formation using a whole-transcriptome approach and the best-studied bacterium, Escherichia coli, and discovered that toxin/antitoxin pairs are related to biofilm formation (Appl. Micro. Biotechnol. 64:515, 2004). Upon further investigation, it was found that the uncharacterized protein YgiU (renamed MqsR) is involved in the regulation of motility through cell-cell signaling that utilizes autoinducer-2 (J. Bacteriol. 188: 305, 2006). We also showed that MqsR functions as a toxin (Nature ISMEJ 2:615, 2008), and the three-dimensional structure of this protein revealed that MqsR functions as an RNase (PLoS Pathogens 5: e1000706, 2009). The antidote to this toxin, MqsA, is an elongated dimer that neutralizes MqsR toxicity as an antitoxin. MqsR adopts an alpha/beta fold that is homologous with the RelE/YoeB family of bacterial ribonuclease toxins. Critically, we have found that MqsA binds promoters of genes important for physiology including mcbR and spy as well as its own promoter through its C-terminal domain. Furthermore, using three sets of whole-transcriptome studies, two nickel-enrichment DNA binding microarrays, and cell survival studies in which MqsR was overproduced in isogenic mutants, we identified eight genes (cspD, clpX, clpP, lon, yfjZ, relB, relE, and hokA) that are involved in MqsR toxicity in addition to its RNase activity (Environ. Micro, in press). Quantitative real-time polymerase chain reaction showed that (i) the MqsR/MqsA complex (and MqsA alone) represses the toxin gene cspD, (ii) MqsR overproduction induces cspD, (iii) stress induces cspD, and (iv) stress fails to induce cspD when MqsR/MqsA are overproduced or when mqsRA is deleted. Hence, upon stress, cells down-regulate their metabolism via toxin MqsR as well as via other toxins controlled by antitoxin MqsA. In addition, we found that deletion of the mqsRA locus decreases persister cell formation (i.e., cells that are resistant to antibiotics without genetic change) and, consistently, overexpression of MqsR increases persister cell formation (BBRC 39:209, 2010); this is the first set of toxin/antitoxin genes found to control persister cell formation. Furthermore, we have found antitoxin MqsA directly controls the transcription of the master regulator of stress, RpoS. Upon stress, MqsA is degraded by proteases which derepresses rpoS. Conversely, production of MqsA represses rpoS and reduces concentrations of the second messenger 3,5-cyclic diguanylic acid due to repression of diguanylate cyclases that are controlled by RpoS. Repression of rpoS by MqsA leads to increased motility and a reduction in the cell adhesins curli and cellulose as well as a reduction in stress resistance and biofilm formation. Hence, external stress alters gene regulation via toxin/antitoxin systems and leads to a switch from the planktonic state (high motility) to the biofilm state (low motility). Therefore, using several systems biology approaches, we have determined the primary roles of toxin/antitoxin systems in cell physiology.

See more of this Session: Genomic Approaches to Systems Biology
See more of this Group/Topical: Topical A: Systems Biology

(281f) Omics Analysis of Agrobacterium Sp. Uncovers Regulatory Factors for Polysaccharide Biosynthesis

Anne Ruffing[1], Marlene Castro[2], Wei-Shou Hu[2] and Rachel Ruizhen Chen[1], (1)Chemical and Biomolecular Engineering, Georgia Institute of Technology, Atlanta, GA, (2)University of Minnesota, Minneapolis, MN

Microbial polysaccharides are primary cellular components and integral to many cellular functions including cell adhesion, biofilm formation, pathogenesis, and protection from the environment. Despite their ubiquity and functional importance, the regulation of polysaccharide biosynthesis remains largely unknown. A polysaccharide-producing *Agrobacterium* sp. was investigated to gain insight into the regulatory factors governing polysaccharide biosynthesis. *Agrobacterium* sp. ATCC 31749 produces a β1,3-glucan polymer known as curdlan, which has applications in a wide-range of industries including food, cosmetics, materials, and medical. While *Agrobacterium* sp. is used industrially to produce curdlan for these commercial applications, very little is known about the genetics and regulatory processes of this microorganism. This study utilizes genomics and transcriptomics to analyze *Agrobacterium* sp. ATCC 31749 at the systems level, revealing a complex, multi-faceted regulatory mechanism for curdlan biosynthesis.

Omics tools provide an avenue for genome-wide analysis of polysaccharide biosynthesis. The genome of ATCC 31749 was sequenced using the Genome Sequencer FLX System from 454 Life Science, yielding a draft genome of 5.4 Mb with 5,585 predicted genes. Subsequently, the annotated draft genome sequence was used to design a custom DNA microarray from Agilent Technologies. Analyzing gene expression levels in *Agrobacterium* sp. identified 985 genes significantly up-regulated and 1,471 significantly down-regulated during curdlan production. The large number of differentially expressed genes can be attributed to the phase change accompanying the initiation of curdlan biosynthesis. Like many other polysaccharides, curdlan is only produced under stress conditions (nitrogen-limitation), when the cell transitions into the stationary phase of the growth cycle. To pinpoint the genes specifically involved in the regulation of curdlan biosynthesis, differentially expressed genes related to three functional categories were selected for further analysis: 1) nitrogen-limited regulation, 2) energy production, and 3) GTP-derived second messengers.

Gene knockout of targets identified from the transcriptome analysis showed four factors to be important for curdlan biosynthesis: the nitrogen signaling cascade, polyphosphate, (p)ppGpp, and c-di-GMP. As expected, the genes of the nitrogen signaling cascade affected curdlan production, but this regulation was shown to be mediated in a rare, RpoN-independent manner. Energy storage in the form of polyphosphate was suspected to be an important energy resource for curdlan production. Gene knockout of polyphosphate-degrading genes confirmed the importance of polyphosphate but also suggested two possible mechanisms of polyphosphate-associated regulation. Lastly, the GTP-derived second messengers, (p)ppGpp and c-di-GMP, were found to affect curdlan biosynthesis and may act as transcriptional, translation, or allosteric regulators. The experimental results of this study will be discussed in light of previously characterized regulatory mechanisms for other microorganisms to provide a global understanding of polysaccharide regulation.

Through using a systems approach to study *Agrobacterium* sp. ATCC 31749, this work uncovered multiple regulatory factors involved in curdlan biosynthesis. Moreover, the genes and factors identified in this study include regulatory elements common to many polysaccharide-producing microbes, and as such, these regulatory factors and mechanisms may be conserved across microbial species. Further investigation of these regulatory factors may be the key to understanding microbial polysaccharide regulation.

See more of this Session: Genomic Approaches to Systems Biology
See more of this Group/Topical: Topical A: Systems Biology

(281g) Automated Metabolic Network Reconstruction: Method, Application and Web Tool

Yu Chen and Xiaoxia Nina Lin, Department of Chemical Engineering, University of Michigan, Ann Arbor, MI

Genome-wide metabolic network reconstruction has proved to be a powerful tool for system-level study of whole-cell metabolisms. A great amount of efforts have been spent in reconstructing metabolic networks of a few tens of organisms, including bacteria, archaea and eukaryotes. However, it remains a time- and labor-consuming task to generate a complete and accurate metabolic network reconstruction that is ready for further applications. On the other hand,a rapidly increasing number of organisms have had or are having their genomes sequenced, which provides first-hand data for whole-cell metabolic network reconstruction. Here, we present a bioinformatics pipeline that can automatically reconstruct metabolic networks of organisms from their genome sequences and annotations. Our bioinformatics pipeline is consists three stages. First, a preliminary metabolic network reconstruction is derived by matching the genes in the genome with the enzymes catalyzing metabolic reactions collected in the Kyoto Encyclopedia of Genes and Genomes (KEGG) database. Second, potential network gaps are predicted by searching for putative metabolic reactions that are required to achieve specific cellular functions, e.g. biomass synthesis. Finally, the probabilities of all the metabolic network gaps are evaluated according to sequence alignment results and the most possible metabolic network are reconstructed by a mixed integer linear programming (MILP) model. In addition, this pipeline fills the network gaps with putative gene candidates in the genome. We first tested this bioinformatics pipeline with Escherichia coli K12 MG1655, the metabolic network of which has been extensively studied. The results from our pipeline are comparable, in terms of the scale of the network, to metabolic network models that were carefully curated manually. Furthermore, we evaluated the phenotype phase plane (PPP) under two conditions using our reconstructed metabolic network, and the results were also in agreement with previous work. Then, the pipeline was applied to twelve strains of Prochlorococcus marinus, the numerically most abundant photosynthetic organism on the earth. By comparing the metabolic reconstructions of different strains of one genus, we can predict the pan and core metabolic networks of Prochlorococcus marinus, which represent the properties of the genus rather than those of individual strains. A publicly available web tool of this bioinformatics pipeline is also under intense development currently. Once established, it will provide services of automated whole-cell metabolic network reconstruction for individual organisms based on genome sequences and annotations.

See more of this Session: Genomic Approaches to Systems Biology
See more of this Group/Topical: Topical A: Systems Biology

(332a) A Comprehensive Profiling of Genome-Wide Transcriptional Programs of Primary Hepatocytes *in Vitro* Cultures

Yeonhee Kim[1], Christopher D. Lasher[2], Logan M. Milford[1], T. M. Murali[3] and Padma Rajagopalan[1], (1)Chemical Engineering, Virginia Polytechnic Institute and State University, Blacksburg, VA, (2) Genetics, Bioinformatics & Computational Biology Program, Virginia Polytechnic Institute and State University, Blacksburg, VA, (3)Computer Science, Virginia Polytechnic Institute and State University, Blacksburg, VA

The liver is one of most important organs in our bodies and performs many essential functions including metabolism, synthesis, secretion, detoxification, and storage. Hepatocytes are the principal cell type that accounts for 80% of liver and involved in major liver-specific functions. In vitro hepatic culture systems are important tools in development of pharmaceuticals, toxicity testing, and the design of bio-artificial liver devices. In hepatic tissue engineering two widely used culture systems are collagen sandwich and hepatocyte monolayers. There has been little investigation into gene expression profiling of hepatocytes for each culture system. Genome-wide gene expression profiles of primary hepatocytes were measured for each cell culture system using Affymetrix GeneChips and analyzed via Gene Set Enrichment Analysis (GSEA) which is a powerful method to elicit biologically meaningful information from microarray data at the level of gene sets.

A comprehensive gene expression study of hepatocytes in collagen sandwich system shows that gene expression monotonically diverges from that in hepatocyte monolayers over an eight-day culture period. Gene sets related to liver-specific functions such as metabolism of cholesterol, fatty acid and amino acid, urea and albumin production, and alcohol metabolism were shown to be gradually up-regulated in the time-dependent manner. Monooxygenases such as cytochrome-P450 enzymes exhibited significant up-regulation in collagen sandwich cultures after three days in comparison to hepatocyte monolayers. Gene sets associated with de-differentiation of hepatocytes were up-regulated in hepatocytes monolayers over the culture period. These temporal genome-wide gene expression studies provide insights into the up- and down-regulation of various liver-specific phenotypic functions and thus help understand subsequent physiological phenomena. These results provide a baseline for further systems biology of engineered liver tissues. Ongoing investigations include the global gene expression profiles of hepatocytes in *in vitro* liver mimics and understanding of underlying inter-cellular communications.

See more of this Session: Genomic Approaches to Systems Biology II
See more of this Group/Topical: Topical A: Systems Biology

(332b) A System Approach to Astrocytoma Progression

Chunjing Wang, Chemical and Biomolecular Engineering, University of Illinois, Urbana-Champaign, Urbana, IL, James A. Eddy, University of Illinois, Urbana, IL and Nathan Price, Department of Chemical and Biomolecular Engineering, University of Illinois, Urbana, IL

Astrocytomas are brain cancers that account for 75% of neuroepithelial tumors. Accurate diagnosis is required for adequate treatment and assessing prognosis for patients. We used both gene-based and network-based approaches to find the key processes involved in astrocytoma progression from pilocytic astrocytoma (WHO grade I) to glioblastoma (WHO grade 4). Out of 20985 genes in Affymetrix 133A array, we identified a core of 494 differentially expressed genes (DEGs) that appeared to be central in astrocytoma. DEGs exhibiting coherent expression profiles were mapped to subnetworks using Expander and enriched GO categories within these subnetworks were identified. On the other hand, our network-based approach identified 193 most perturbed expressed pathways that are essential to astrotytoma pathophysiology. These pathways were characterized into different biological processes involving regulation of immune system process, regulation of signal transduction, regulation of MAPK activity, regulation of cell cycle and cell death, among others. A majority of these processes showed deregulation, while immune system related pathways became more tightened as the disease became more aggressive. Patients with glioblastoma, the most aggressive form of astrocytoma, showed heterogenity in their gene expression profiles and we identified a pathway, MTA3 which could effectively classify GBM patients into two subtypes with significant differences ($p<0.05$) in survival time. Our systems approach provides a comprehensive basis for developing models for astrocytomas and suggests possible therapeutic approaches.

See more of this Session: Genomic Approaches to Systems Biology II
See more of this Group/Topical: Topical A: Systems Biology

(332c) Chronic Alcohol Effects On the System-Wide Transcriptional Regulatory Network Modules and Dynamics During Rat Liver Regeneration

Rajanikanth Vadigepalli, Biswanath Patra, Egle Juskeviciute, Harikrishnadhar Sonnenahalli, Sara Crumm and Jan B. Hoek, Department of Pathology, Anatomy and Cell Biology, Thomas Jefferson University, Philadelphia, PA

Liver regeneration is a complex, yet remarkably coordinated, process characterized by system-wide gene expression changes underlying the response of the liver tissue to acute damage through mechanical, chemical or viral mechanisms. This normal regenerative response is impaired by acute or chronic alcohol exposure. Such an alcohol-mediated disruption of repair mechanisms presumably contributes to the onset of liver damage in alcoholic liver disease. While progress has been made in understanding several aspects of liver regeneration, the system-wide mechanisms by which these are impaired by chronic alcohol exposure remain largely unexplained. This study aims to characterize how chronic ethanol treatment affects the transcriptomic regulatory dynamics in regenerating liver.

The overall driving biological hypothesis of the present project is that the disruptive effects of chronic alcohol on the liver regeneration involve alterations in a focused set of TFs acting in a network to regulate system-wide downstream target genes. We take an integrated experimental and computational approach to investigate the dynamics of the system-wide transcriptomic changes, use sensitive bioinformatics methods to infer regulatory modules comprising of transcription factors (TFs) and target genes, and validate these for changes in the activity of TFs and their binding at target gene promoters.

In our experimental approach, rats were fed a liquid diet containing 36% of total calories derived from ethanol; corresponding pair-fed calorie-matched controls received liquid diets in which ethanol calories were replaced either by carbohydrate or by fat. After five weeks, rats were subjected to 2/3d PHx by surgical excision of left-lateral and medial (LLM) lobes. Remnant liver tissues were harvested after 1, 6 and 24h to represent the priming phase, early G1, and G1/S transition, respectively. LLM tissues obtained at t=0 served as within-animal controls. We used Affymetrix Rat Gene 1.0 ST arrays with ~25000 probe sets to obtain global gene expression data from each liver sample (n=4 replicate animals, 72 arrays total).

Analysis of the gene expression data obtained from the LLM tissue samples (12 replicates per diet group) revealed a total of approximately 400 differentially expressed genes (>1.5-fold change) in ethanol-adapted animals compared to either control diet. No significant differences between the high-fat and high-carbohydrate control groups were obtained. Functional annotation analysis using DAVID

Bioinformatics resource (v. 6.7) indicated significant alterations in gene clusters related to circadian rhythm, lipid and steroid metabolism, and ER membrane-related functions. The effect of PHx, analyzed using a 3-way ANOVA, revealed a total of approximately 5000 transcript clusters as responsive to PHx in at least one of the diet groups at either 1, 6 or 24h post PHx (false discovery rate $< 15\%$). Hierarchical clustering yielded >35 distinct dynamic gene expression response profiles, several with altered response pattern in ethanol fed rats compared to those fed carbohydrate or fat control diets. Within the control and post PHx clusters, alcoholic livers showed distinct expression profiles and clustered separately from those of the carbohydrate and high fat fed rats.

In each of the gene expression clusters, over-represented transcriptional regulatory elements were identified in corresponding promoters using our PAINT bioinformatics software, indicating that chronic ethanol intake significantly affects the transcriptional regulatory network dynamics in response to PHx. We experimentally validated the effects of chronic alcohol exposure on 16 TFs actively involved in liver regeneration. Our results reveal that ethanol exposed rats showed: (1) suppression of FosB, JunD and JunB, GATA-1, C/EBP-α, PPAR-α, PPAR-γ, NRF2 and ATF2, (2) delayed activation of NF-kB, HNF1, STAT3, p-CREB and C/EBP-β, (3) transient activation of c-Fos and c-Jun. We applied a novel regulatory module identification approach that was previously employed on genome-wide location analysis data to our PAINT bioinformatics predicted TF binding sites. Our analysis revealed that these TFs are likely to operate in combinatorial modules with co-localized regulatory elements.

We predicted NF-kB binding at 21 target promoters using our PAINT bioinformatics approach and validated these predictions using Chromatin Immunoprecipitation assay. Our results reveal a baseline activation of NF-kB in alcoholic livers, further increase in binding activity at 1h post PHx for all promoters but increase at 6h post PHx occurring at only a subset of the target genes. We propose that the baseline increase in NF-kB activity due to chronic alcohol intake limits further changes in promoter binding at key genes, and this may underlie the adverse affects of chronic alcohol treatment on liver regeneration.

Taken together, our results provide a first system-wide view of the regulatory network dynamics in liver regeneration as it is affected by alcohol, and implicate combinatorial regulatory modules as participating in the adverse liver regeneration response.

Supported by: AA008714, AA014986, AA017261, AA016919, and AA018873.

See more of this Session: Genomic Approaches to Systems Biology II
See more of this Group/Topical: Topical A: Systems Biology

(332d) Genomic Approaches to Evolutionary History of Primate Cytochromes c

Lawrence I. Grossman[1], Denis Pierron[1], Juan C. Opazo[1], Margit Heiske[2], Zack Papper[3], Monica Uddin[4], Gopi C. Markondapatnaikuni[1], Derek E. Wildman[3], Roberto Romero[5] and Morris Goodman[6], (1)Center for Molecular Medicine and Genetics, Wayne State University, Detroit, MI, (2)Laboratoire de Physiopathologie Mitochondriale, Université Victor Segalen Bordeaux 2, Bordeaux, France, (3)Center for Molecular Medicine and Genetics, Wayne State University School of Medicine, Detroit, MI, (4) Epidemiology, University of Michigan, Ann Arbor, MI, (5)Eunice Kennedy Shriver National Institute of Child Health and Human Development, NICHD, NIH, DHHS, Detroit, MI, (6)Anatomy & Cell Biology and Center for Molecular Medicine and Genetics, Wayne State University, Detroit, MI

Cytochrome c (cyt c) is a highly conserved protein among eukaryotes, mainly owing to its pivotal role in two crucial cellular processes: apoptosis and respiratory chain function. Despite its overall stability, previous studies have reported on the human lineage: (i) a high amino acid evolution rate of cyt c somatic isoform, (ii) absence of the testis isoform, and (iii) atypical biochemical behavior of human cyt c. In order to gain insight, we have investigated the cyt c loci and sequence evolution among primate lineages. Using cyt c sequences obtained by sequencing and from databases, genomic loci were obtained and compared using the UCSC Genome Browser. Phylogenetic statistical approaches were used to construct trees, to estimate divergence times, and to test selection models. Then potential effects of the changes were evaluated with spatial comparison and mathematical modeling. All the analyzed primate testis cyt c sequences share the same nonsense mutation, which suggests that the silencing occurred in the early primate stem. The phylogenetic analyses of somatic cyt c converged to the same tree topology recovering all major groups with maximal branch support. The evolutionary analyses show that strong positive selection occurs specifically on the anthropoid lineage root and then continued in parallel on the early catarrhini and platyrrhini stem. Spatial analysis and mathematical modeling suggest that evolution specifically focus on the respiratory chain rather than apoptosis or other cyt c functions. Supported by previous biochemical studies, our results show that the silencing of the cyt c testis isoform is correlated with the decrease of primate reproduction and that the fast evolution of the cyt c somatic isoform is correlated with the increase of primate metabolism. Finally, the parallel evolution of cyt c in the two sister anthropoid groups leads us to propose that cyt evolution may in fact have participated in the evolutionary events with which its changes are correlated.

See more of this Session: Genomic Approaches to Systems Biology II
See more of this Group/Topical: Topical A: Systems Biology

(332e) Multi-Platform Genome Re-Sequencing of Evolved Strains of the Cellulolytic Actinobacterium Thermobifida Fusca

Yu Deng, Chemical and life science engineering, Virginia Commonwealth University, Richmond, VA and Stephen S, Fong, Virginia Commonwealth University, Richmond, VA

Currently, one of the hurdles hindering efficient production of cellulosic biofuel is the recalcitrant nature of cellulose to hydrolysis. Thermobifida fusca is a high G-C content, thermophilic, gram-positive soil actinobacterium with high cellulolytic activity. T. fusca was adaptively evolved on the cellobiose and on glucose/cellobiose which generated a stress-relief cycle for evolution and then two strains: a specialist phenotype muC and a generalist phenotype muS that both converged to a similar cell yield (1.5g/L) were gained respectively. Both strains appeared to adjust their growth phenotype by modulating cellulase activity and intracellular ATP levels. To characterize the mutations of the genome during the evolutions, the high-throughput Illumina and 454 sequencing technologies were used. By re-sequencing these two strains by Illumina and 454 sequencing, 18 mutations in T. fusca muC and 14 mutations in T. fusca muS were confirmed as well as 48 found to be errors in the reference genome sequence of wild-type T. fusca. The genome-scale constraint-based model of T. fusca was built and verified by experiment. To connect the mutations caused by adaptive evolution to the metabolic changes, the flux variability of mutant and wild-type were predicted by the model and compared with each other.

See more of this Session: Genomic Approaches to Systems Biology II
See more of this Group/Topical: Topical A: Systems Biology

(332f) Towards Genome-Scale Metabolic Reconstruction of Plant Metabolism

Rajib Saha, Patrick F. Suthers and Costas D. Maranas, Department of Chemical Engineering, The Pennsylvania State University, University Park, PA

The scope and breadth of genome-scale metabolic reconstructions has continued to evolve over the last decade. However, only a limited number of efforts to reconstruct plant metabolism are currently available and all of them are focused on a single model plant system (i.e., Arabidopsis thaliana). In this talk, we present progress in the development of a genome-scale model for a plant with direct applications to foodstuff and bioenergy production (i.e., maize). Maize is a monocot grass species used as a model plant to study and understand plant biology and genetics. Unlike A. thaliana, annotation of maize is still underway which introduces significant challenges in the association of metabolic functions to genes. The developed model is designed to meet rigorous standards on gene-protein-reaction (GPR) associations, elementally and charged balanced reactions and a biomass reaction showing how different biomass precursors are involved into the existence of a plant cell. The metabolic network contains over 1,800 metabolites involved in over 1,600 reactions from primary and secondary metabolism of maize. As many as 350 reactions and 200 metabolites are unique to maize compared to A. thaliana. All reactions are elementally and charged balanced and localized into five different intracellular compartments (i.e., cytoplasm, mitochondrion, plastid, peroxisome and vacuole). GPR associations are also established based on the functional annotation information and homology prediction and categorized into five major classes of monofunctional proteins, multifunctional proteins, isozymes, multimeric proteins and protein complexes. We describe results from performing flux balance analysis under different physiological conditions, (i.e., photosynthesis, photorespiration and respiration) and explore their impact on biomass formation, lignin formation, and carbohydrate yield. The developed model corresponds to the largest and more complete to-date effort at cataloguing metabolism for a plant species.

See more of this Session: Genomic Approaches to Systems Biology II
See more of this Group/Topical: Topical A: Systems Biology

(332g) Liver Disease Diagnosis Based On Gene Expression Profiling

Yuliang Wang[1], Jaeyun Sung[1], Shuyi Ma[1] and Nathan Price[2], (1)Chemical and Biomolecular Engineering, UIUC, Urbana, IL, (2)Department of Chemical and Biomolecular Engineering, University of Illinois, Urbana, IL

Major liver diseases such as hepatocellular carcinoma, cirrhosis, and hepatitis affect huge numbers of people. For example, each year, death from liver cancer alone is 662,000 worldwide. Early diagnosis of liver disease is crucial. If identified in time, hepatitis C patients can be treated with interferon to avoid cirrhosis; and cirrhosis patients can be closely monitored and treated to stop progression to liver cancer. Previous gene expression analysis of liver diseases mainly focuses on identifying transcriptomic signatures that differentiate a single liver disease (such as hepatocellular carcinoma) from normal liver. Disease signatures identified in this way may not be specific to a particular liver disease, but may be a common signature for multiple liver pathological states because other liver diseases are considered. In our study, we collected microarray data of over 400 liver biopsy samples from Gene Expression Omnibus, representing multiple liver diseases such as hepatocellular carcinoma , cirrhosis with or without concomitant hepatocellular carcinoma , liver dysplastic nodule, and hepatitis C. Then we used a Top Scoring Pair based decision tree algorithm to classify all these liver diseases simultaneously and found a unique transcriptomic signature for each liver disease. The algorithm finds gene pairs that show relative expression reversal among different liver diseases and builds the decision tree in a bottom up way. Classifying all these major liver diseases simultaneously ensures that the identified disease signatures truly represent the unique properties of each liver disease, because each signature is identified against all major liver pathological states, not just normal liver. The identified unique transcriptomic signatures may provide valuable information for targeted biomarker discovery for liver disease.

See more of this Session: Genomic Approaches to Systems Biology II
See more of this Group/Topical: Topical A: Systems Biology

(404b) Study of Yeast Sphingolipid Metabolism with a Detailed Kinetic Model

Paolo Angelino and **Vassily Hatzimanikatis**, Institute of Chemical Sciences and Engineering, Ecole Polytechnique Fédérale de Lausanne (EPFL), Lausanne, Switzerland

Sphingolipids have recently emerged as important bioactive molecules in addition to being critical structural components of cellular membranes. Beside their structural function, these molecules have been implicated in regulating cell growth and fundamental cell processes such as differentiation, migration, and apoptosis. On the organism level, sphingolipids play roles in physiological processes including inflammation and vasculogenesis. We developed a kinetic model of the sphingolipid metabolism in yeast with the aim to identify the critical key parameters responsible for the regulation of the sphingolipid metabolism and its interaction with the global cell metabolism. The mechanistic model allows for an accurate description of the kinetics of each reaction and of enzyme saturation. In particular, we have investigated the effects of the competition of multiple substrates for a single enzyme on the distribution of fluxes and on the steady state concentrations. Sphingolipids in yeast are present in five hydroxylation states, which have been shown to interact differently with other lipids, as for example ergosterol, to form cell membrane structures. Our model describes the hydroxylation states and it is able to reproduce their experimental profiles. We performed a detailed sensitivity analysis of the steady-state concentrations and fluxes and we identified the key parameters that determine the function of the sphingolipid pathway and the distribution of the hydroxylation states. We further discuss how the model can be extended to include larger portions of the lipid and carbon biosynthesis pathways. Knowledge acquired from the yeast model can be readily applied to other organisms, since the spingolipid pathways is highly conserved in eukaryotes.

See more of this Session: In Silico Systems Biology: Cellular and Organismal Models
See more of this Group/Topical: Topical A: Systems Biology

(404c) AUMIC: An Automated Tool for Developing Dynamic Metabolic Models

Hyun-Seob Song and Dorwaiswami Ramkrishna, School of Chemical Engineering, Purdue University, West Lafayette, IN

We introduce an automated tool for developing dynamic metabolic models. A software is developed using MATLAB and named as AUMIC which stands for AUtomated tool for Metabolic modeling Integrated with the Cybernetic regulatory mechanisms. While AUMIC is designed mainly for constructing cybernetic models including lumped cybernetic model (LCM) (e.g., Kompala et al., 1986), hybrid cybernetic model (HCM) (Kim et al., 2008; Song et al., 2009), and lumped hybrid cybernetic model (L-HCM) (Song and Ramkrishna, 2010), it can also handle various other approaches to dynamic metabolic modeling such as lumped kinetic model (LKM) (e.g., Bijkerk and Hall, 1977), macroscopic bioreaction model (MBM) (Provost and Bastin, 2004), and dynamic flux balance analysis (dFBA) (Mahadevan et al., 2002) for comparative studies. LKM, MBM and dFBA can be considered as respective counterparts of LCM, HCM and L-HCM because of analogous features in each pair. In AUMIC, all of the foregoing approaches are subsumed as quasi-steady-state models which are formulated through consistent procedures within a unifying framework, i.e., network decomposition into elementary modes (EMs) using METATOOL (von Kamp and Schuster, 2006), EM classification and processing, and design of kinetics for uptake fluxes (and kinetics for enzyme synthesis). Moreover, AUMIC provides the parameter identification routine so that one can develop their own models using specific experimental data. Finally, the results are quickly analyzed using the post-processing module. Conveniently, all tasks in AUMIC are performed using graphical user interface. Continuing efforts are being made towards developing AUMIC as an even more effective and advanced tool and its first version will be released soon for public use.

References:

Bijkerk AHE, Hall RJ. 1977. Mechanistic Model of Aerobic Growth of Saccharomyces-Cerevisiae. Biotechnology and Bioengineering 19(2):267-296

Kim JI, Varner JD, Ramkrishna D. 2008. A Hybrid Model of Anaerobic E. coli GJT001: Combination of Elementary Flux Modes and Cybernetic Variables. Biotechnology Progress 24(5):993-1006

Kompala DS, Ramkrishna D, Jansen NB, Tsao GT. 1986. Investigation of Bacterial-Growth on Mixed Substrates - Experimental Evaluation of Cybernetic Models. Biotechnology and Bioengineering 28 (7):1044-1055

Mahadevan R, Edwards JS, Doyle FJ. 2002. Dynamic flux balance analysis of diauxic growth in

Escherichia coli. Biophysical Journal 83(3):1331-1340 Provost A, Bastin G. 2004. Dynamic metabolic modelling under the balanced growth condition. Journal of Process Control 14(7):717-728.

Song HS, Morgan JA, Ramkrishna D. 2009. Systematic Development of Hybrid Cybernetic Models: Application to Recombinant Yeast Co-consuming Glucose and Xylose, Biotechnology and Bioengineering 103(5): 984-102

Song HS, Ramkrishna D. 2010. Prediction of Metabolic Function from Limited Data: Lumped Hybrid Cybernetic Modeling (L-HCM). Biotechnology and Bioengineering 106(2): 271-284

von Kamp A, Schuster S. 2006. Metatool 5.0: fast and flexible elementary modes analysis. Bioinformatics 22(15):1930-1931

See more of this Session: In Silico Systems Biology: Cellular and Organismal Models
See more of this Group/Topical: Topical A: Systems Biology

(404d) Probabilistic Integrative Modeling of Genome-Scale Metabolic and Regulatory Networks in Escherichia Coli and Mycobacterium Tuberculosis

Sriram Chandrasekaran, Biophysics and Computational Biology, Institute for Genomic Biology, University of Illinois, Urbana-Champaign, Urbana, IL and Nathan, D. Price, Center for Biophysics and Computational Biology, Department of Chemical and Biomolecular Engineering,, Institute for Genomic Biology, University of Illinois, Urbana-Champaign, Urbana, IL

Systems biology seeks to understand the cell by studying interactions between its various components and the properties that emerge from these interaction networks. The integration of diverse network types has several important applications including diagnosing metabolic disorders and discovering novel drug targets. A cardinal challenge in this process is the integration of transcriptional regulatory networks with the corresponding metabolic network. We propose a new method called **Probabilistic Regulation of Metabolism (PROM)** which achieves this by performing **automated integrated modeling** of regulatory and metabolic networks using high throughput data. Using PROM, we construct an integrated regulatory-metabolic network for the model organism, *Escherichia coli*, and demonstrate PROM's ability to model the effect of perturbations to transcriptional regulators and predict microbial growth. After validating the approach, we then applied the method to build the **first genome-scale** integrated metabolic–regulatory model for ***Mycobacterium tuberculosis***, a critically important human pathogen. This study incorporates data from over 1300 microarrays, 2000 transcription factor–target interactions regulating 3300 metabolic reactions, and 1905 knockout phenotypes for E. coli and M. tuberculosis. PROM has the ability to identify drug targets and was able to predict the growth phenotypes with accuracies as high as 90%. Importantly, this method represents the successful integration of two important classes of systems biology models that are rarely combined quantitatively.

See more of this Session: In Silico Systems Biology: Cellular and Organismal Models
See more of this Group/Topical: Topical A: Systems Biology

(404e) Metabolic Reconstruction and Analysis of Methane Producing Methanosarcina Barkeri

Matthew C. Gonnerman, Chbe, UIUC, Urbana, IL and Nathan, D. Price, Center for Biophysics and Computational Biology, Department of Chemical and Biomolecular Engineering,, Institute for Genomic Biology, University of Illinois, Urbana-Champaign, Urbana, IL

Methane is a clean-burning renewable resource that is often overlooked as a potential bioenergy source. Methane already has a well established infrastructure and many uses ranging from fuel for certain forms of transportation to heating homes. Methanosarcina barkeri has the ability to produce methane through the process of methanogenesis. Methanogenic Archaea grow on a limited substrate profile including methanol, hydrogen with carbon dioxide and acetic acid. The methanosarcina species can be manipulated to consume a wider variety of substrates including byproducts and inhibitors of current fermentation processes. Starting from a metabolic reconstruction of Methanosarcina barkeri, we have updated the model under the guidance of experienced methanogenic archea researchers. The current reconstruction links the previous draft annotation with the current gene annotation as well as incorporates new thermodynamic estimations utilizing published algorithms (Group Contribution Method). The reconstruction consists of 470 enzymatic genes that govern 764 reactions. Constraint-based analysis of the metabolic model offers guidance when incorporating new substrate pathways as well as a method to align natural objectives (growth) with engineering objectives (methane production). We can perform simulated gene insertions and deletions in the model and make predictions of experimental outcomes. The model serves as a guide for current experimentation in M. barkeri to increase production of methane.

See more of this Session: In Silico Systems Biology: Cellular and Organismal Models
See more of this Group/Topical: Topical A: Systems Biology

(404f) The Genome-Scale Model Biomass Constituting Equation Plays a Large Role in Predicting Biofuels Production

Ryan S. Senger, Biological Systems Engineering Department, Virginia Tech, Blacksburg, VA

Genome-scale models have made important contributions to biotechnology and biomedicine. When used with flux balance analysis for predictive purposes, maximizing the specific growth rate is the common objective function. To simulate cell growth, a biomass constituting equation is required and consists of several macromolecules (e.g., protein, DNA, RNA, lipids, etc.) as well as organic metabolites, energy, reducing power, and ionic species. Sensitivity analyses have been performed on different biomass constituting equations of several genome-scale models. In most cases, relatively low influence of the stoichiometric coefficients of biomass components on the resulting specific growth rate has been observed. Because of this, it is easy to interpret that the biomass constituting equation holds little significance to genome-scale modeling when compared to the catabolic pathways. Here, this assumption is shown to be false as the predicted production of biobutanol from *Clostridium acetobutylicum* ATCC 824 is shown to be a strong function of the biomass constituting equation. First, several existing biomass constituting equations (from closely-related organisms) were evaluated using an existing *C. acetobutylicum* model to determine the influence of the biomass constituting equation on the relationship between the specific growth rate of the organism and the predicted biobutanol production. Next, the specific proton influx/efflux rate showed to have a strong influence on this relationship, as did cellular maintenance requirements. The relative amounts of specific lipids in the biomass constituting equation showed to have little effect, demonstrating a need for solvent toxicity mechanisms to be built into genome-scale models. These findings have several implications. In particular, it is demonstrated that the biomass constituting equation should no longer exist as a static entity in a genome-scale model, and steps are presented on how to develop dynamic biomass equations. The concept of a global sensitivity analysis of the biomass constituting equation is presented here, and the implications of this approach are contrasted with the relative sensitivity analysis approaches that have been used to date.

See more of this Session: In Silico Systems Biology: Cellular and Organismal Models
See more of this Group/Topical: Topical A: Systems Biology

(468a) The Influence of Circadian Rhythms On the Inflammatory Response

Jeremy D. Scheff[1], Steven E. Calvano[2], Stephen F. Lowry[2] and Ioannis P. Androulakis[3], (1)Biomedical Engineering, Rutgers University, Piscataway, NJ, (2)Department of Surgery, UMDNJ-Robert Wood Johnson Medical School, New Brunswick, NJ, (3)Biomedical Engineering, Chemical & Biochemical Engineering, Rutgers University, Piscataway, NJ

The inflammatory response is a key component in the defense against a variety of harmful stimuli, such as an invading pathogen or trauma. Inflammation, consisting of a complex, coordinated set of interactions between the immune system and the neuroendocrine system, initiates the restoration of homeostasis either through the removal of the pathogen or the repair of damaged tissue. Typically, inflammation is tightly regulated, activating when necessary and returning to homeostasis after healing has been initiated. However, inflammation does not always resolve appropriately; in some cases, an elevated level of inflammation persists, which can damage otherwise healthy tissue. Prolonged systemic inflammation comes with severe consequences, often leading to organ failure and death. This type of overwhelming inflammatory when accompanied by an infection is called sepsis. There are approximately 750,000 cases of severe sepsis every year in the United States alone, leading to over 200,000 deaths annually (Angus, Linde-Zwirble et al. 2001). Thus, the management of inflammation is a major challenge in the treatment of critically ill patients.

Despite our understanding of the importance of this problem and extensive research towards the development of effective therapies, current treatment options are not adequate and the development of new treatments has been slow (Freeman and Natanson 2000). This may be due to the inherent challenges in applying reductionist techniques to complex, nonlinear systems (Seely and Christou 2000). It may not be possible to predict the outcome of perturbing a pathway involved in inflammation if knowledge of its behavior comes only from its study in isolation (Vodovotz, Clermont et al. 2004). This has stimulated interest in the development of quantitative, systems biology models of inflammation.

In recent years, a number of models have been developed by applying different modeling techniques (agent based modeling or equation based modeling), at different scales (molecular, cellular, systemic, or a combination), and focusing on different specific problems (acute inflammation, trauma, or the response to a specific disease) (Vodovotz, Constantine et al. 2009). These models have been developed with the practical goals of impacting healthcare through translational systems biology and rationalizing the design of experiments and clinical trials (Clermont, Bartels et al. 2004). Because of the large number of components involved in inflammation, existing models make assumptions about which interactions are most important, either by simplifying or neglecting certain elements. One aspect that has not previously been studied from the perspective of systems biology is the interplay between circadian rhythms and inflammation.

Circadian rhythms are biochemical, behavioral, or physiological processes that are entrained to a 24 hour periodic cycle. This rhythmicity is widely observed in humans from the scale of biochemical reactions, such as hormone production, to behavioral patterns, such as regular sleeping and feeding times. In the context of healthcare, mouse and rat models have shown that the same dose of a drug can be lethal at certain times and ineffective at others (Levi and Schibler 2007). Thus, it is not surprising that there is also a circadian component to inflammation; in fact, many of the elements typically included in models of inflammation (leukocytes, cytokines, and hormones) are known to have strong diurnal patterns (Coogan and Wyse 2008). The importance of these variations is apparent by observing that sepsis patients have a heightened risk of mortality between 2am and 6am (Hrushesky, Langevin et al. 1994).

Many of the components included in previous attempts at modeling inflammation have been shown to exhibit circadian patterns. Several studies have shown that numerous pro- and anti-inflammatory cytokines undergo diurnal variations in plasma levels, typically peaking in the night (Zabel, Horst et al. 1990; Petrovsky and Harrison 1997; Petrovsky and Harrison 1998; Petrovsky, McNair et al. 1998; Hermann, von Aulock et al. 2006). Plasma cortisol levels also exhibit a circadian pattern, peaking in the early morning. Cortisol is produced by the actions of the hypothalamic-pituitary-adrenal axis, and the circadian production is due to stimulation from the central circadian clock in the suprachiasmatic nucleus (SCN).

Due to the immunomodulatory effects of glucocorticoids and the strong circadian pattern of plasma cortisol levels, cortisol has been implicated in the circadian entrainment of cytokine production (Petrovsky and Harrison 1998). However, glucocorticoids to not affect all cytokines equally; it stimulates the production of anti-inflammatory cytokines while inhibiting the production of pro-inflammatory cytokines (Barber, Coyle et al. 1993; Barnes 1998). Thus, since most cytokines have peak levels around the same time, it seems unlikely that cortisol alone could be responsible for the observed fluctuations in cytokine level, especially in light of the fact that a number of other hormones also vary either in or out of phase with cytokine levels (Petrovsky and Harrison 1998).

One of these hormones that has received particular attention is melatonin, which is important due to its potential role as a mediator in the crosstalk between the SCN and the immune system (Coogan and Wyse 2008). Melatonin is tightly regulated to have a peak in production in the night while remaining at very low levels the rest of the day and it has been shown to stimulate the production of cytokines (Guerrero and Reiter 2002; Skwarlo-Sonta, Majewski et al. 2003). This is supported by experimental evidence showing that pinealectomy leads to decreased cytokine production in mice (Delgobbo, Libri et al. 1989). Thus melatonin is used as the primary circadian regulator of cytokine production in this model. Melatonin and cortisol drive the circadian variation in all of the other model variables. These interactions are used to develop a model that includes the normal diurnal variations in model components as well as time-of-day-dependent responses to inflammatory stimuli. This work has applications in translational systems biology due to the importance of considering circadian rhythms when treating inflammatory diseases (Hrushesky and Wood 1997).

References

Angus, D. C., W. T. Linde-Zwirble, et al. (2001). "Epidemiology of severe sepsis in the United States: Analysis of incidence, outcome, and associated costs of care." Critical Care Medicine 29(7): 1303-1310.

Barber, A. E., S. M. Coyle, et al. (1993). "Glucocorticoid Therapy Alters Hormonal and Cytokine Responses to Endotoxin in Man." Journal of Immunology 150(5): 1999-2006.

Barnes, P. J. (1998). "Anti-inflammatory actions of glucocorticoids: molecular mechanisms." Clinical Science 94(6): 557-572.

Clermont, G., J. Bartels, et al. (2004). "In silico design of clinical trials: A method coming of age." Critical Care Medicine 32(10): 2061-2070.

Coogan, A. N. and C. A. Wyse (2008). "Neuroimmunology of the circadian clock." Brain Research 1232: 104-112.

Delgobbo, V., V. Libri, et al. (1989). "Pinealectomy Inhibits Interleukin-2 Production and Natural-Killer Activity in Mice." International Journal of Immunopharmacology 11(5): 567-573.

Foteinou, P. T., S. E. Calvano, et al. (2009). "Translational Potential of Systems-Based Models of Inflammation." Clinical and Translational Science 2(1): 85-89.

Freeman, B. D. and C. Natanson (2000). "Anti-inflammatory therapies in sepsis and septic shock." Expert Opinion on Investigational Drugs 9(7): 1651-1663.

Guerrero, J. M. and R. J. Reiter (2002). "Melatonin-immune system relationships." Current Topics in Medicinal Chemistry 2(2): 167-179.

Hermann, C., S. von Aulock, et al. (2006). "Endogenous cortisol determines the circadian rhythm of lipopolysaccharide- but not lipoteichoic acid-inducible cytokine release." European Journal of Immunology 36(2): 371-379.

Hrushesky, W. J. and P. A. Wood (1997). "Circadian time structure of septic shock: timing is everything." J Infect Dis 175(5): 1283-4.

Hrushesky, W. J. M., T. Langevin, et al. (1994). "Circadian Dynamics of Tumor-Necrosis-Factor-Alpha (Cachectin) Lethality." Journal of Experimental Medicine 180(3): 1059-1065.

Levi, F. and U. Schibler (2007). "Circadian rhythms: Mechanisms and therapeutic implications." Annual Review of Pharmacology and Toxicology 47: 593-628.

Petrovsky, N. and L. C. Harrison (1997). "Diurnal rhythmicity of human cytokine production - A dynamic disequilibrium in T helper cell type 1/T helper cell type 2 balance?" Journal of Immunology 158 (11): 5163-5168.

Petrovsky, N. and L. C. Harrison (1998). "The chronobiology of human cytokine production." International Reviews of Immunology 16(5-6): 635-649.

Petrovsky, N., P. McNair, et al. (1998). "Diurnal rhythms of pro-inflammatory cytokines: Regulation by plasma cortisol and therapeutic implications." Cytokine 10(4): 307-312.

Seely, A. J. E. and N. V. Christou (2000). "Multiple organ dysfunction syndrome: Exploring the paradigm of complex nonlinear systems." Critical Care Medicine 28(7): 2193-2200.

Skwarlo-Sonta, K., P. Majewski, et al. (2003). "Bidirectional communication between the pineal gland and the immune system." Canadian Journal of Physiology and Pharmacology 81(4): 342-349.

Vodovotz, Y., G. Clermont, et al. (2004). "Mathematical models of the acute inflammatory response." Current Opinion in Critical Care 10(5): 383-390.

Vodovotz, Y., G. Constantine, et al. (2009). "Mechanistic simulations of inflammation: current state and future prospects." Math Biosci 217(1): 1-10.

Vodovotz, Y., M. Csete, et al. (2008). "Translational systems biology of inflammation." Plos Computational Biology 4(4).

Zabel, P., H. J. Horst, et al. (1990). "Circadian rhythm of interleukin-1 production of monocytes and the influence of endogenous and exogenous glucocorticoids in man." Klinische Wochenschrift 68(24): 1217-1221.

See more of this Session: In Silico Systems Biology: Cellular and Organismal Models II
See more of this Group/Topical: Topical A: Systems Biology

(468b) In Silico Modeling of the Immune Response Network: Influenza Immune-Escape Strategies and Implications in Influenza Drug Research

Jason E. Shoemaker[1], Yukiko Matsuoka[1], Hiromi Matsumae[1], Satoshi Fukuyama[1], Ken Fuji[1], Yukiko Muramoto[1], Hiroaki Kitano[2] and Yoshihiro Kawaoka[1], (1)Japanse Institute of Science and Technology, Tokyo, Japan, (2)Systems Biology Institute, Tokyo, Japan

Recent evidence suggests that an overly ambitious immune response may be the primary factor that determines influenza severity [1], yet the definition of what is a reasonable immune response remains vague. Seasonal influenza, alone, is responsible for approximately 30,000 deaths in the US each year. With the recent swine origin influenza virus (SOIV) pandemic and the fear that highly pathogenic avian influenza strains (H5N1) could attain human-human transmissibility, there is a growing interest in identifying host-viral interactions. A network understanding of the viral-host interaction will greatly facilitate the identification of immune escape strategies employed by influenza viruses. Furthermore, these *in silico* models may be employed to identify drug targets which may selectively alter viral replication capacity or mediate collateral damage from an aberrant immune response [2]. To date, we have created models of viral replication in epithelial cells and inter-cellular communication between lymphocytes against which microarray and protein data can be interpreted. Applying mRNA and protein data to these models, down-regulation of the chemokines responsible for directing lymphocyte chemotaxis has been identified as a potential immune-escape strategy for highly-pathogenic influenzas.

In silico models of viral replication and immune signaling were constructed to predict the chemokines released from infected cells and the impending effect of the chemokines on the recruitment and chemotaxis of lymphocyte cells, respectively. Currently, the viral replication consists of approximately 200 proteins and 320 host-viral, protein-protein/protein-RNA interactions. Upon infection, the virion fuses with the cell membrane, resulting in transportation of the viral ribonucleoprotein (vRNP) complex to the nucleus where viral replication begins. While several intra-cellular, host defense pathways are circumvented by the viral proteins, infection still results in the production of chemokines. These chemokines act as the input into the immune signaling model and are responsible for the migration of lymphocyte cells (T/B cells, natural killer (NK) cells) as well as directing lymphocytes to the site of infection. Chemokines emanating from infected epithelial cells attract NK cells and macrophages. As macrophages approach the site of infection, they release additional chemokines which attract T and B cells. In an ideal immune-response situation, this communication results in the recognition and destruction (via apoptosis) of viral-infected cells. The viral replication model has been constructed using CellDesigner, and equations are being developed to better explain to complex behaviors observed in immune-related, intra-cellular signaling. The immune signaling model currently consists of 8 states in which transport to the site of infection is modeled as different states.

To explore differences in the immune response to particular influenza strains, mRNA expression and protein levels were measured in bronchial epithelial swabs and sera, respectively, for monkeys exposed to a seasonal influenza, the SOIV, and an avian H5N1 strain. Data was collected from 12 monkeys at days 1, 3, 5, and 7 post infection (p.i.). Clustering of the microarray data, interpreted using GO analysis, showed that the behavior of key biological process, such as cell cycle and RNA production, hijacked by the viruses during infection, is highly conserved between the strains. Furthermore, immune related genes are up regulated day 1 p.i. for all three strains. Protein data in the blood also showed little difference between the infection types with one notable exception. Key interleukins known to increase immunoglobin secretion could not be detected in SOIV samples. This may explain the why SOIV can achieve higher viral loads that seasonal influenza in lung tissue.

Microarray data reveals a possible immune-escape mechanism for highly pathogenic influenza infections that's physiological effects can be correlated with the immune signaling model. At days 5 and 7 p.i., the H5N1 infected data shows a sudden down-regulation of many of the immune signaling genes. Interpreted through the immune signaling model, the effect of this observed regulation would be the continued induction of lymphocyte recruitment but a strong inhibition of the coordinated chemotaxis required to bring the lymphocytes to the site of infection. As viral replication continues, large populations of lymphocytes can accumulate, leading to apoptosis and inflammation in the tissue without abating the infection. This prediction from the model is consistent with several experimental studies [3] and histopathological studies are underway to provide direct observational evidence.

The models of the intra and inter-cellular immune response provide several advantages to drug research and development. As more viral-host interactions become known, it is hoped that pathway redundancy may allow for the identification of host pathways which are critical for viral replication yet temporarily dispensable for the host. Furthermore, drug regimen strategies which mediate the immune-response instead of the viral-replication mechanism can be explored. Should similar immune-escape strategies be observed in other highly pathogenic strains, these drug regimens may proof to be more robust than many modern treatments as the virus itself is not being targeted.

In conclusion, *in silico* models of the immune response provide an invaluable platform for elucidating viral infection strategies and for drug target exploration/strategy development. Currently, several projects in our lab focus on many aspects of modeling the immune response. Data-driven and literature-driven methods of protein-protein interaction network elucidation along with protein pull-down essays are being used to explore for additional host-viral interactions. Furthermore, better modeling platforms are being designed that can account for both biochemical modeling and mechanical motion (endocytosis and chemotaxis). As predictive capacity increases, these models will provide valuable insights into the workings of the immune response and the immune-escape strategies of other viral/bacterial infections.

1. Peiris JS, Cheung CY, Leung CY, Nicholls JM: **Innate immune responses to influenza A H5N1: friend or foe?** *Trends Immunol* 2009, **30**(12):574-584.

2. Kobasa D, Jones SM, Shinya K, Kash JC, Copps J, Ebihara H, Hatta Y, Kim JH, Halfmann P, Hatta M *et al*: **Aberrant innate immune response in lethal infection of macaques with the 1918 influenza virus**. *Nature* 2007, **445**(7125):319-323.

3. Rowe T, Leon AJ, Crevar CJ, Carter DM, Xu L, Ran L, Fang Y, Cameron CM, Cameron MJ, Banner D *et al*: **Modeling host responses in ferrets during A/California/07/2009 influenza infection**. *Virology* 2010.

See more of this Session: In Silico Systems Biology: Cellular and Organismal Models II
See more of this Group/Topical: Topical A: Systems Biology

(468c) Cell Population Heterogeneity and Mass Transport Dynamics Modulate Tissue Regeneration Processes

Pauline A. Markenscoff[1], Jian Feng[1] and **Kyriacos Zygourakis**[2], (1)Electrical and Computer Engineering, University of Houston, Houston, TX, (2)Chemical and Biomolecular Engineering, Rice University, Houston, TX

A better understanding of the complex interplay between cell population dynamics and mass transport is necessary for overcoming the barriers that have slowed down the development of bioartificial tissues.

In a recent publication [1], we presented the development of a multi-scale computational model that describes the dynamic behavior of homogeneous cell populations growing under mass transfer limitations in three-dimensional scaffolds. The model has three components: a transient partial differential equation for the simultaneous diffusion and consumption of a limiting nutrient or growth factor; a cellular automaton describing cell migration, proliferation and collisions in a cubic lattice; and equations that quantify how the varying nutrient or growth factor concentration modulates cell division and migration. The hybrid (discrete-continuous) model was parallelized and solved on a distributed-memory multicomputer to study how transport limitations affect tissue regeneration rates under conditions encountered in typical bioreactors. Simulation results revealed that the severity of transport limitations can be estimated by the magnitude of two dimensionless groups: the Thiele modulus and the Biot number. The initial spatial distribution of seed cells, their migration speed and the hydrodynamics of the bioreactor also influenced the overall rate and the pattern of tissue growth.

This study presents a significant extension of this model to heterogeneous cell populations growing in three-dimensional scaffolds with "artificial" vessels. To account for different cell phenotypes, the new model considers cell populations exhibiting bimodal distributions of migration speeds, persistence and division times. The model also allows for scaffold heterogeneities, cell differentiation, natural cell death (apoptosis) and cell death caused by nutrient or growth factor depletion (necrosis).

Simulation of three-dimensional tissue growth is a computationally challenging problem requiring large grids to capture the growth of tissues of practically significant size. To meet this challenge, our algorithm has been parallelized to run on a Cray XD1 computer cluster. The computational domain is partitioned into subdomains along one dimension (slab decomposition) and each subdomain is assigned to a different processor.

Our simulation results reveal that cell differentiation or necrosis can lead to the formation of complex tissue architectures when heterogeneous cell populations grow under conditions leading to severe transport limitations. These tissue patterns can also be affected by the initial spatial distribution of seed cells, time-dependent addition of the critical nutrient or growth factor to the media and the hydrodynamics of the bioreactor that modulate the boundary conditions of the diffusion-reaction problem.

Finally, we explore the effect of vascularization on the growth rates and structure of regenerated tissues. "Artificial" vessels with cylindrical or other shapes are placed in the 3D scaffolds to enhance the transport of nutrients to the interior of developing tissues. We will present simulation results that show how tissue development is affected by (a) the geometry of the artificial vessels, and (b) temporal variations in the flow rates of culture media or the concentrations of limiting nutrients.

These results provide us with guidelines for a rational design of *in vitro* cell culture experiments that will lead to the formation of bioartificial tissues with desired architecture.

Reference:

[1] G. Cheng, P. Markenscoff and K. Zygourakis, "A 3D Hybrid Model for Tissue Growth: The Interplay Between Cell Population and Mass Transport Dynamics", *Biophysical Journal*, **97**, 401-414 (2009).

See more of this Session: In Silico Systems Biology: Cellular and Organismal Models II
See more of this Group/Topical: Topical A: Systems Biology

(468d) Modeling and Analysis of the Retinoic Acid Induced Proliferation and Differentiation Program of HL-60

Ryan Tasseff[1], Satyaprakash Nayak[1], Sang Ok Song[1], Andrew Yen[2] and Jeffrey D. Varner[1], (1) Chemical and Biomolecular Engineering, Cornell University, Ithaca, NY, (2)Biomedical Sciences, Cornell University, Ithaca, NY

Understanding the molecular basis of differentiation programs is one of the grand unmet challenges facing modern cell-biology. In this study, we integrated computational and experimental network analysis to unravel the response of HL-60 human myeloblastic leukemia cells to Retinoic Acid (RA) stimulation. HL-60, because of its defined but limited ability to differentiate, is an archetype model to study the molecular architecture of human differentiation programs. HL-60 differentiates down two possible lineages depending upon external stimuli such as Retinoic Acid (RA) or Vitamin D3 (D3). Our initial studies have focused on the role of the BLR1 receptor in the transduction of RA signals. BLR1, a Gq-protein coupled receptor expressed following RA exposure, is required for RA-induced cell-cycle arrest and differentiation and leads to atypical persistent MAPK signaling. A dynamic mathematical model of RA induced cell-cycle arrest and differentiation was formulated and tested against BLR1 wild-type (wt), knock-out and knock-in HL-60 cell-lines with and without RA. The current HL-60 model architecture described the dynamics of 729 proteins and protein complexes interconnected by 1356 interactions. An ensemble strategy was used to compensate for uncertain model parameters. Consistent with previous experimental studies, the initial HL-60 model showed up-regulation of BLR1 expression following RA exposure along with sustained MAPK activation. The model also captured the feedback between BLR1 expression and MAPK activation in BLR1 knock-out and Raf-knock-down cell-lines. Lastly, the model predicted the direction of RA-induced expression shifts for a panel of proteins not used in model training. Taken together, our modeling efforts have established a prototype organization for the differentiation program in HL-60. BLR1 acts as a component of a feed-forward control element which drives its own expression and that of other proteins required for differentiation. The initial simulation studies have also led to testable structural linkages between BLR1 expression and MAPK activation that we are exploring experimentally. More broadly, we have demonstrated that modeling of molecular programs can prioritize experimental directions and expand current biological knowledge despite model uncertainty.

See more of this Session: In Silico Systems Biology: Cellular and Organismal Models II
See more of this Group/Topical: Topical A: Systems Biology

(468e) Exploration of Signaling Cycles Using Dynamic Optimization

Andrijana Radivojevic[1], Benoît Chachuat[2], Dominique Bonvin[1] and Vassily Hatzimanikatis[3], (1) Automatic Control Laboratory, Ecole Polytechnique Federale de Lausanne (EPFL), Lausanne, Switzerland, (2)Department of Chemical Engineering, McMaster University, Hamilton, ON, Canada, (3) Institute of Chemical Sciences and Engineering, Ecole Polytechnique Fédérale de Lausanne (EPFL), Lausanne, Switzerland

One of the basic characteristics of every living system is the ability to respond to extracellular signals. This is carried out through a limited number of protein-based signaling networks, whose function is not only based on simple transmission of the received signals, but incorporates the processing, encoding and integration of both external and internal signals. The results than lead to different changes in gene expression, regulate cell growth, mitogenesis, differentiation, embryo development, and stress responses in mammalian cells, whereas the malfunction is in correlation with diseases.

Commonly observed instances of signal transduction through a series of protein kinase reactions are mitogen activated protein kinase (MAPK) cascades. These pathways are found in almost all eukaryotes and play an important role in controlling different cellular processes, including fundamental functions.

In order to better understand the nature of this regulation and to gain greater insight into the mechanisms that determine the function of cells, MAPK cascades have been intensively studied using mathematical modeling and computational simulations. The primary aim is to faithfully describe the system and to be able to predict the system behavior. Synergistically with experimental analysis, reported observations have identified properties of these pathways, such as rapid induction, noise resistance, amplification capability, threshold induction mechanism, resistance to "cross-talk", etc.

Here, we investigate one class of approaches for analyzing the relationship between network structure and functional behavior and the overall idea involves applying optimization techniques. By manipulating the desired functional behavior and by monitoring the corresponding parameter values, one can learn how model parameters and functions are related, and then be in a position to discover new design principles.

The primary motivation was to explore if there is any trade-off while promoting simultaneously large amplification and fast signal propagation. We identified the competing parameters in the linear tricyclic cascade and their values for the optimal design leading to minimal response times and given amplification. We also incorporated "ultrasensitivity", in order to analyze interplay between this steady-state property and the dynamic behavior of the system. Special emphasis is placed on the robustness of the resulting tricyclic cascades in the face of variations in kinase and phosphatase concentration ratios.

(468f) Stochastic Simulations of the Tetracycline Operon

Konstantinos Biliouris, Yiannis Kaznessis and Prodromos Daoutidis, Chemical Engineering and Materials Science, University of Minnesota, Minneapolis, MN

The tetracycline (tet) operon is a robust self-regulated system. It is found naturally in bacteria where it confers resistance to antibiotic tetracycline. Because of the performance of the molecular elements (promoters, operators, repressor) of the tetracycline operon, these elements are widely used as parts of synthetic gene networks. More specifically, in many synthetic systems the expression of desired proteins is under the transcriptional control of the tet components, where protein production can be efficiently turned on and off in response to the presence or the absence of tetracycline.

The present work examines the mechanisms that govern the dynamics of the tetracycline operon. We have formulated a mathematical model of the naturally occuring tetracycline operon guided by experimental findings. This model incorporates the biomolecular interactions of this system, including those involved in transcription, translation, degradation, protein dimerization, repression and induction. The model represents each interaction with biochemical reactions.

A hybrid, stochastic-discrete and stochastic-continuous algorithm was used to simulate the tetracycline operon behavior. A sensitivity analysis with respect to important parameters underlying this system was also performed, providing mechanistic insight into the working of this interesting system. The results of the simulations are in agreement with, and explain well experimental observations such as tight repression, fast gene expression, induction with tetracycline, and small intracellular repressor TetR2 amounts. The simulations determine the exact amounts of intracellular TetR2 and TetA protein. The former is the repressor molecule whereas the latter mediates the removal of tetracycline from the cell. Furthermore, the results demonstrate that one of the two promoters of the tetracycline operon is redundant and not functionally important in Escherichia coli. Sensitivity analysis illustrates that changes in the affinity of tetracycline for the repressor TetR2 and of TetR2 for the operator sites have a significant impact on the behavior of the tetracycline operon, suggesting optimum interaction strengths developed through natural selection.

This work affords augmented insight into the interplay between the molecular components of the tetracycline operon. It also provides useful explanations of how the components and their interactions have evolved to best serve the bacteria carrying this operon. Thus, it may assist in designing new antibiotics that circumvent the existing resistance mechanism. It can also aid the design of novel, yet superior synthetic gene networks.

See more of this Session: In Silico Systems Biology: Cellular and Organismal Models II

(541a) Gene Regulation by MAPK Substrate Competition

Yoosik Kim, Chemical Engineering, Princeton University, Princeton, NJ, Ze'ev Paroush, The Hebrew University-Hadassah Medical School, Jerusalem, Israel, Gerardo Jimenez, Institut de Biologia Molecular de Barcelona-CSIC, Barcelona, Spain and Stanislav Y. Shvartsman, Princeton University, Princeton, NJ

Introduction

Traditional view of pattern formation in the early Drosophila embryo is essentially one-dimensional: The anterior, posterior, and terminal systems initiate the formation of bands and stripes of gene expression along the AP axis; at the same time, the orthogonal gradient of Dorsal (Dl) controls a different set of genes along the DV axis. These maternal patterning signals were believed to act independently from one another until they integrate at the enhancer of their zygotic target genes [1]. Recently, we have shown that in addition to this widely accepted idea of a genetic developmental program, the maternal signals can integrate upstream of the zygotic gene regulation through a substrate competition mechanism [2]. Specifically, the anterior morphogen Bicoid (Bcd) interacts with the terminal system and acts as a competitive inhibitor of the terminal signaling.

In the early embryo, the terminal patterning is mediated by localized activation of Mitogen Activated Protein Kinase/Extracellular Regulated Kinase (MAPK/ERK). MAPK functions downstream of Torso, a uniformly expressed receptor tyrosine kinase that is presented with its ligand at the anterior and posterior poles of the embryo. Spatially restricted Torso signaling regulates the expression of genes required for the development of the terminal structures such as tailless and huckebein. This depends on MAPK-mediated phosphorylation of transcriptional repressors Capicua (Cic) and Groucho (Gro), both of which are uniformly distributed throughout the embryo [3].

In the anterior region, Bcd competes with Cic and Gro and other MAPK substrates to bind to the activated MAPK (dpERK), lowering MAPK mediated downregulation of these repressors. As a consequence, the level of Cic is significantly higher in the anterior region compared to the posterior [2]. However, the functional significance of this asymmetric MAPK signaling at the termini remains unclear. Using a biophysical model and genetic experiments, we propose that this observed asymmetry of MAPK signaling is functionally significant and can explain genes that exhibit both AP and DV polarity. Specifically, we suggest that wild type expression of zerknullt (zen) requires this differential signaling gradient of the terminal system. Our work suggests an additional layer of integration of AP and DV signaling systems and provides a plausible mechanism for regulation of 2-dimensional gene expression patterns in the embryo.

Results and Discussion

Our mathematical model is based on the quantitative experimental information about Bcd, Dl, and terminal gradients, and biophysical descriptions of reaction and diffusion in the syncytial blastoderm. Since Dl gradient acts perpendicular to Bcd and dpERK, the embryo needs to be modeled as a 2-D system. Although the real embryo looks like an ellipsoid, for illustrative purposes, we considered a rectangular embryo where the x and y directions represent AP and DV axes of the real embryo respectively. We assumed that the AP and DV systems do not interact until they integrate at the DNA to regulate their common target gene such as zen: Hence the two axes are treated as two independent systems.

Our model analyzes the regulatory interactions of zen described previously [4]. Briefly, zen is activated by a spatially uniform activator (X), but is repressed by Dl and repressor complex (Dl-R) [4]. Since zen repression requires both Dl and R, we model their joint action by the product of their concentrations. At the poles, dpERK antagonizes Dl mediated repression as it deactivates Dl corepressor through phosphorylation. In the anterior region, Bcd presents as an extra substrate of dpERK that can compete with R. Bcd gradient is modeled as an exponentially decaying curve with maximum at the anterior tip (x=0). The competition between Bcd and the R acts as if Bcd is an indirect inhibitor of dpERK mediated phosphorylation of R. We use the switch-like model of gene regulation, where zen is only expressed when the product Dl and R is below some constant threshold, To. Thus, the level of zen expression is given by zen(x,y) = H(To-R(x)*Dl(y)), where H(u) is the Heaviside step function: H(u) = 1, when u>0 and 0 otherwise.

The spatial pattern of the repressor, R(x), is found from a reaction-diffusion problem that describes the MAPK-mediated phosphorylation of the repressors. We assume that the MAPK kinase (A), its phosphatase (D), and Bcd are in excess; hence, their concentrations profiles are parameters/inputs and not variables of the model. The spatial pattern of MAPK kinase, A(x) activity has peaks at both the anterior and the posterior poles. We neglect the dephosphorylation of MAPK substrates, and assume that all species have the same diffusivity and all substrates are phosphorylated with the same kinetics. Under these assumptions, the model has 6 variables: unphosphorylated repressor (R), inactive and active MAPK (E and E*), and active MAPK in complex with Bcd, R, or MAPK phosphatase (BE*, RE*, and DE*, respectively). The model predicts that the level of R is higher at the anterior than the posterior pole. This reflects the effect of substrate competition since Bcd is distributed in the AP gradient and thus, competes with the repressor for binding to phosphorylated MAPK only in the anterior region, resulting in an AP asymmetric pattern of R.

A level set (line of constant value) of the product of the Dl and R patterns, Dl(y)*R(x), gives the 2D boundary of the zen pattern. Our model can predict how this pattern changes in response to variations in the levels of anterior, terminal, and DV signals. For instance, the model correctly predicts that the ectopic expression of zen in embryos that lack Dl. In addition, the terminal signaling has much stronger effect on antagonizing Dl mediated repression at the posterior pole. With regard to the asymmetric pattern of zen, our model predicts that lowering the dose of Bcd should lead to ectopic zen at the anterior since MAPK signaling activity toward the uniform repressor is increased in the absence of its competing substrate, Bcd.

To test this prediction, we examined zen expression in embryos derived from bcd null mothers. We found that zen is no longer repressed in the anterior region in these embryos. In the absence of Bcd, zen is expressed all around anterior pole and shows nearly symmetric pattern at both poles. This expansion most likely reflects increase level of MAPK signaling in these embryos since Dl gradient is not significantly affected in these embryos, and zen is expressed in the anterior region despite the high level of nuclear Dl. In the wild type embryos, anterior zen is expressed only to the point where Dl level becomes very low, whereas at the posterior pole, zen and Dl are co-expressed. However, in the embryos lacking Bcd, zen and Dl are co-expressed not only at the posterior, but also at the anterior pole.

Conclusion

We have developed a network type interaction model for regulation of zen. Specifically, we have focused on the AP asymmetry of zen expression, reflecting differential modulation by the terminal signaling at the poles. Based on the biophysical model, we argue that zen is repressed in the anterior region due to stronger MAPK substrate competition induced by Bcd.

Our results provide new insights to coupling of different signaling systems in patterning of the early embryo. We have established that MAPK can integrate inputs from other maternal morphogens such as Bcd via substrate competition. Through a biophysical model and genetic experiments, we showed that this new mechanism of signal integration is functionally significant; it is required for patterning of zen. In the future, our model can be extended to examine the joint effects of anterior, terminal, and dorsoventral signaling systems in patterning of the embryo.

References

[1] Papatsenko, D., Goltsev, Y., and Levine, M. (2009) Organization of developmental enhancers in the Drosophila embryo. Nucleic Acid Research 37, 5665-5667.

[2] Kim, Y., Coppey, M., Grossman, R., Ajuria L, Jimenez, G., Paroush, Z., and Shvartsman, SY. (2010). MAPK substrate competition integrates patterning signals in the Drosophila embryo. Current Biology. 20, 446-451.

[3] Jimenez, G., Guichet, A., Ephrussi, A., and Casanova, J. (2000). Relief of gene repression by Torso RTK signaling: role of capicua in Drosophila terminal and dorsoventral patterning. Genes & Development 14, 224-231.

[4] Rusch, J. and Levine, M. (1994) Regulation of the dorsal morphogen by the Toll and torso signaling pathways: a receptor tyrosine kinase selectively masks transcriptional repression. Genes & Development 8, 1247-1257.

(541b) Quantitative Modeling and Analysis of TGF-β-Induced Erk/MAPK and Smad Signaling

Seung-Wook Chung[1], Mary C. Farach-Carson[2], Carlton R. Cooper[3] and Babatunde A. Ogunnaike[1], (1) Chemical Engineering, University of Delaware, Newark, DE, (2)Biochemistry and Cell Biology, Rice University, Houston, TX, (3)Biological Sciences, University of Delaware, Newark, DE

Transforming growth factor β (TGF-β), which plays a pivotal role in regulating a wide array of cellular and physiologic processes in both normal and pathological cells, is known to function as a potent tumor suppressor during the early stage of tumorigenesis, primarily by inhibiting cell proliferation and inducing apoptosis. However, in the late phases of tumor progression, the role of TGF-β appears to become one of tumor promotion, apparently supporting growth, subverting the immune system, and also facilitating epithelial-to-mesenchymal transition (EMT), invasion and angiogenesis. While these various cellular responses are known to be induced via several TGF-β-driven signaling cascades that include both the canonical Smad and non-Smad pathways, significant research efforts have been devoted almost exclusively to the canonical Smad-meditated intracellular signaling pathway due to its primary anti-proliferative effects; the non-Smad pathways, are less well-characterized, even though they also play non-trivial roles in carcinogenesis. Of the various non-Smad pathways of interest, a growing body of evidence has demonstrated the importance of Erk/MAPK signaling during cancer progression because TGF-β-induced Erk activation is critical for EMT, a process that ultimately leads to tumor cell invasion and metastasis. To obtain a more complete quantitative understanding of the role of TGF-β signaling during cancer progression, we have developed a comprehensive mathematical model of TGF-β signaling that combines the Erk-MAPK pathways with the canonical Smad pathway using information from experimental observations reported in the literature. By describing how an extracellular TGF-β ligand signal is sensed by receptors and transmitted via these two separate intracellular pathways, the model yields quantitative insight into how TGF-β-induced responses can be modulated and regulated. The model also predicts possible dynamic behavior of the Erk-mediated pathway in cancerous cells that provide clues regarding possible explanatory mechanisms for the seemingly contradictory roles of TGF-β during cancer progression. In particular, a comparison of the differences in the dynamics of transcriptionally active Smad and Erk between normal and cancerous signaling systems suggests that an imbalance between lipid-raft and non-raft TGF-β receptor internalization can lead to TGF-β-induced EMT in cancer cells.

See more of this Session: In Silico Systems Biology: Intracellular Signaling and Gene Regulation
See more of this Group/Topical: Topical A: Systems Biology

(541c) Interferon-Mediated Innate Immune Responses to Virus Infection

Musarat Ishaq, Chemical and Biological Engineering, University of Wisconsin-Madison, Madison, WI and John Yin, Chemical and Biological Enginering, University of Wisconsin-Madison, Madison, WI

Vesicular Stomatitis Virus (VSV) is an important vaccine vector and also has potential applications as an oncolytic (tumor-killing) virus. Strains of VSV that stimulate a type I interferon (IFN) response are an important hallmark in the design of RNA-virus based therapeutic strategies. VSV wild type can shutdown its host innate immune response. VSV matrix (M) protein blocks the cellular antiviral response by interrupting the cellular transcription machinery and blockading nucleocytoplasmic mRNA transport through cellular nucleopore protein 98 (Nup98). Here, we used different VSV mutants in which the N gene, encoding the viral nucleocapside protein), was moved from its 3` promoter proximal position in the wild-type virus (N1), toward the 5` distal end, defining mutants N2, N3 and N4. We also studied M protein mutant (M51R) in which Methionine (M) 51 is converted to Arginine(A), where the mutation inactivates M-protein-mediated blocking of the cellular antiviral response. We found that VSV-N3 induced several genes involved in innate immune response with a similar profile to VSV-M51R; both strains induced the expression of key transcriptional factors like STAT1. In our global study of gene expression by RNA microarray, we observed broad patterns of differential gene expression that reflected differences in viral strains: genes 2 fold up regulated (N1≥1617, N2≥344, N3≥627, N4≥50, M51R≥891) and 2 fold down regulated (N1≥3839, N2≥223, N3≥558, N4≥46 and M51R≥741). VSV transcription is sequential process under the control of single 3` transcription initiation site, where polymerase enter and transcribe viral mRNAs as a gene order N>P>M>G>L. Surprisingly, the expression of VSV-G reflects a different pattern in all mutants (N>G>P>M>L),. These studies begin to highlight how subtle and not-so-subtle changes in the timing and level of viral gene expression can impact the pattern of host defensive responses.

See more of this Session: In Silico Systems Biology: Intracellular Signaling and Gene Regulation
See more of this Group/Topical: Topical A: Systems Biology

(541d) Identification of Families of Signal Transduction Models Using Pareto Optimal Ensemble Techniques (POETs)

Anirikh Chakrabarti, Sang Ok Song and Jeffrey D. Varner, Chemical and Biomolecular Engineering, Cornell University, Ithaca, NY

Mathematical modeling of complex signal transduction and gene expression programs is an emerging tool for understanding disease mechanisms. However, conventional wisdom suggests that the data requirement to identify and validate complex mechanistic models is too large. Typically, it is not possible to uniquely identify model parameters, even with extensive training data and perfect models. This reality has brought into the foreground a number of interesting questions. For example, do we actually need exact parameter knowledge to predict qualitatively important properties of a molecular network? Ensemble approaches, which use parametrically and potentially even structurally uncertain model families, have emerged to deal with uncertainty in systems biology and other fields like weather prediction. Their central value has been the ability to quantify simulation uncertainty and to constrain model predictions, despite sometimes only order-of-magnitude parameter estimates. In this study, we introduce Pareto Optimal Ensemble Techniques (POETs) to identify a family of simplified proof-of-concept signal transduction models. POETs integrate Simulated Annealing (SA) with Pareto optimality to estimate parameter sets on or near the optimal tradeoff surface between competing training objectives. The proof-of-concept model described in a general way the integration of extracellular signals with kinase activation, the phosphorylation of transcription factors and the up-regulation of an associated transcriptional program (including molecular descriptions of transcription, translation initiation and nuclear transport). Thus, while the example was not specific to a particular growth factor, signal integration cascade or expression program, it contained many of the general features encountered when identifying specific models. We modeled the molecular interactions in the prototypical signaling network using mass action kinetics within an ordinary differential equation (ODE) framework (64-ODEs in total). We assumed spatial homogeneity but differentiated between cytosolic, membrane and nuclear localized processes. The true model was used to generate synthetic data sets from which we tested the ability of POET algorithm to identify the 117 unknown model parameters. Each synthetic measurement was assumed to be a Northern or Western blot. To this end, we implemented a novel scaling procedure that allowed the systematic integration of these types of measurements into the model identification problem. Thus, we knew only relative amounts of protein or mRNA for any specific condition or time. To constrain the absolute concentration scale, we assumed a single ELISA or qRT-PCR measurement for the highest intensity band in each case. Lastly, we limited our training data to 20 samples per experiment (an upper limit on the lanes available on a Western blot).

The POET algorithm generated an ensemble of 1600 signal transduction models with several interesting features. First, the ODE model used here was deterministic and did not describe stochastic gene expression fluctuations. However, because many many different models were sampled, the deterministic ensemble exhibited population-like behavior. For example, scaled gene expression levels were approximately normally distributed following the addition of extracellular ligand. Thus, while gene expression was not described at a single-cell level, the ensemble captured coarse-grained expression heterogeneity. This suggested that deterministic ensembles could perhaps be used to model heterogeneous populations when stochastic simulation is not tractable. Second, the model ensemble captured critical robustness and fragility features of the true model, despite significant parameter uncertainty. Edge and node ranks computed over the ensemble recovered the true rankings for highly fragile and highly robust network components. This suggested that, in practice, results from sensitivity analysis obtained by analyzing model ensembles could represent true behavior, at least for highly fragile or robust network features. Taken together, analysis of the experimentally constrained ensemble of models generated using POET suggested that we don't need exact parameter information to understand qualitatively important network features.

See more of this Session: In Silico Systems Biology: Intracellular Signaling and Gene Regulation
See more of this Group/Topical: Topical A: Systems Biology

(541e) Integration of Signaling Networks with Differential Gene Regulation Data Using a Probabilistic Pathway Impact Analysis Approach

Albert Babinskas and **Rajanikanth Vadigepalli**, Department of Pathology, Anatomy and Cell Biology, Thomas Jefferson University, Philadelphia, PA

The objective of the present study is to integrate the information on intracellular signaling network structures with large-scale gene expression data sets to predict functional changes in the cellular pathways that are operational under those conditions. Conventional approaches do not consider the network structure and transform the signaling pathway structures into lists of proteins, map the corresponding gene expression to these proteins and compute statistical confidence for observing these changes against a random chance. As the network structure is ignored in such 'enrichment'-based methods, the results do not reflect the biological constraints and are likely to be erroneous in typical use. For example, consider the case when only a single component of a pathway with 10 total nodes is significantly down-regulated. The biological significance is likely to be dependent on whether the down-regulated component is an activator or an inhibitor and its location in the pathway structure. However, conventional methods ignore this important connectivity information. Recently, a method named SPIA was proposed to overcome this issue by incorporating the signaling pathway structure explicitly in assessing the impact of gene expression changes on cellular function (Traca et al., Bioinformatics 2009). A limitation of SPIA is that it considers the canonical signaling pathways in its assessment and does not reflect the particular biological context from which the gene expression data was obtained.

We developed an alternative approach to SPIA that identifies context-specific signaling network structures that are consistent with the observed gene expression data. Our approach starts with a user-specified pathway structure, generates thousands of alternative pathway structures and evaluates each network for consistency with the gene expression data based on the probability of perturbation computed by the SPIA algorithm. This approach allows us to remove those interactions in the initial signaling pathway that are not consistent with the changes at the gene expression level. The probability of perturbation of each network and the enrichment-based statistics are combined into a significance metric for statistical evaluation of each network.

We benchmarked our approach using three biologically motivated synthetic networks. In the benchmark studies, we considered 5-node, 7-node and 11-node synthetic networks with multiple activation and inhibitory interactions and feedback connections. In parallel, we generated nominal gene expression data sets with up-regulated activators and down-regulated inhibitors (data sampled from a normal distribution with appropriately signed mean of +/-1 and variance 0.09). For each of the benchmark cases, we searched for networks that are consistent with the simulated gene expression data sets by removing one or more connections in the pathways. Our results indicate that (1) networks with high consistency are likely to arise by differential expression of activators and inhibitors in opposing directions, (2) quantitative changes in the nodes with multiple upstream activation and inhibitory connections significantly affect the consistency, and (3) nodes with feedback interactions have large influence on whether the network is consistent with the gene expression data.

In the biological case study, we applied the validated method to evaluate the biologically relevant interactions in the Angiotensin II receptor type 1 (AT1R) mediated pathway operational in the rat brainstem neurons dysregulated under high blood pressure conditions. We considered microarray gene expression data obtained from hypertensive animals and normotensive controls. The initial AT1R network was based on literature-based pathway interactions. We explored sixteen connections in the pathway as candidates for removal from the network structure and examined each of the 2^16 (=65536) networks for consistency with the microarray gene expression data using the probability of perturbation determined by SPIA. The networks with highly significant consistency showed remarkable similarity within the treatment groups but are distinct for hypertensive vs normotensive samples. Based on these results, we propose that the AT1R pathway interactions in the brainstem neurons are conditionally dependent on the hypertensive state of the animal.

In summary, we present a novel approach to integrate signaling pathway structure information with the differential gene regulation and derive context-specific network structures based on the available gene expression data.

Research Support: NIH R01GM083108 and R33HL088283.

See more of this Session: In Silico Systems Biology: Intracellular Signaling and Gene Regulation
See more of this Group/Topical: Topical A: Systems Biology

(541f) Caveats of Parametric Sensitivity Analysis: Dynamical Analysis of Biological Systems

Thanneer Malai Perumal, Chemical and Biomolecular Engineering, National University of Singapore, Singapore, Singapore and **Rudiyanto Gunawan**, Department of Chemical and Biomolecular Engineering, National University of Singapore, Singapore, Singapore

The challenge of understanding complex cellular networks has spurred the use of quantitative (mathematical) modeling and analysis in biology [1]. These quantitative models represent virtual biochemistry, mimicking biochemical reactions in the cell [2]. Here, sensitivity analysis, a well established method in science and engineering literature [3], is the most frequently applied analysis due to its ease of computation and intuitive interpretation of the results. In systems biology, this analysis has become a powerful tool to investigate the manner of which model outputs or behavior depend on model parametrization [4], and has also been included in majority of off-the-shelf systems biology tools [5-8]. This analysis has been used for many purposes, such as model calibration and model identifiably, model validation and reduction, finding bottle-necking processes of the system [4].

There are generally two types of sensitivity analysis that are employed in biological modeling; local and global. Local sensitivity analysis concerns with the change in the system behaviour with respect to an infinitesimal change in the parameter from its nominal value. On the other hand, global sensitivities describe the effect of simultaneous or large variations of parameters on the system behavior [3]. Local analysis is used more often than global as it is easier to compute and interpret. The computed sensitivity coefficients give information regarding the contribution of individual parameter to the system output behavior. The magnitude of the sensitivities thus indicates the degree of importance, based on which parameter ranking can be generated.

Many biological behaviors are characterized by their dynamic response, for example in signal transduction pathways and gene regulatory networks. In this case, ODE models are often used to simulate such dynamics using time-dependent state or output variables (such as concentrations). The model parameters typically include physicochemical constants (such as those related to reaction kinetics, transport properties, etc.) as well as initial conditions, operating conditions, and geometric parameters of the systems. The physicochemical parameters are measured experimentally or estimated theoretically or chosen ad-hoc, and therefore have uncertainties. In addition, the initial and operating conditions may change in time for a variety of reasons, such as varying receptor concentration and environmental conditions like temperature. Parametric sensitivity coefficients are computed as the ratio between the differences in state variable caused by changes in the parameter values and used to investigate the impact of parameter uncertainty on model outputs.

Some examples of applications of parametric sensitivity analysis in dynamic biological models include programmed cell death [9], budding yeast cell cycle control [10], IL-6 signaling pathway [11], circadian rhythm model in Neurospora [12], coupled MAPK and PI3K signal transduction pathway [13]. The interpretations of the resulting time varying parametric sensitivities depend on the applications. In most cases, consolidated sensitivity metrics, e.g. using integrals of sensitivity magnitudes or absolute sum of sensitivities over time or the Fisher Information Matrix, are used to indicate parametric importance and to generate parameter ranking. In doing so, such metrics can mislead the modeler in understanding system dynamics. In addition, one should be cautious in inferring dynamical information from time-dependent parametric sensitivities since parametric perturbations are made on the system parameters. As these parameters are usually constants and do not have dynamics, the parametric sensitivity coefficients thus reflect an integrated effect of parameter change on the system behavior.

In this work, we show that parametric sensitivity analysis, either taken directly or as consolidated metrics, can give wrong conclusions about the bottle-necking or governing process in a dynamical system. Two examples, an artificial network and a Fas-induced programmed cell death model [14] are used to illustrate this issue. In addition, we offer a modified parametric sensitivity analyses (both global and local) that can reveal dynamical importance of parameters directly using modified local sensitivities (local analysis) and information transfer (global analysis). Application to the cell death model above revealed the importance of mitochondrial-dependent pathway, while standard parametric sensitivities pointed to the direct caspase-8 pathway as the most sensitive. Comparison of this result to pathway knock-outs in silico confirmed the control of cell death regulation by mitochondria in this model. This work highlights the pitfalls of using parametric sensitivity analysis to infer dynamical bottle-necking process.

References:

1. Kitano, H., Systems biology: a brief overview. Science, 2002. 295(5560): p. 1662--1664.

2. Szallasi, Z., J. Stelling, and V. Periwal, System Modeling in Cell Biology From Concepts to Nuts and Bolts. 2006, Cambridge, Massachusetts: The MIT Press.

3. Varma, A., M. Morbidelli, and H. Wu, Parametirc Sensitivity in Chemical Systems, ed. A. Varma, M. Morbidelli, and H. Wu. 1999: Cambridge University Press, Cambridge, UK.

4. Ingalls, B., Sensitivity analysis: from model parameters to system behaviour. Essays Biochem, 2008. 45: p. 177-93.

5. Maiwald, T. and J. Timmer, Dynamical modeling and multi-experiment fitting with PottersWheel. Bioinformatics, 2008. 24(18): p. 2037-43.

6. Hoare, A., D.G. Regan, and D.P. Wilson, Sampling and sensitivity analyses tools (SaSAT) for computational modelling. Theor Biol Med Model, 2008. 5: p. 4.

7. Klipp, E., et al., Systems biology standards--the community speaks. Nat Biotechnol, 2007. 25(4): p.

390-1.

8. Zi, Z., et al., SBML-SAT: a systems biology markup language (SBML) based sensitivity analysis tool. BMC Bioinformatics, 2008. 9: p. 342.

9. Shoemaker, J.E. and F.J. Doyle III, Identifying Fragilities in Biochemical Networks: Robust Performance Analysis of Fas Signaling-Induced Apoptosis. Biophys J, 2008.

10. A. Lovrics, I.G.Z., A. Csikász-Nagy, J. Zádor, T. Turányi, B. Novák,, Analysis of a budding yeast cell cycle model using the shapes of local sensitivity functions. International Journal of Chemical Kinetics, 2008. 40(11): p. 710-720.

11. Chu, Y., A. Jayaraman, and J. Hahn, Parameter sensitivity analysis of IL-6 signalling pathways. IET Syst Biol, 2007. 1(6): p. 342-52.

12. Jin, Y., et al., Uniform design-based sensitivity analysis of circadian rhythm model in Neurospora. Computers and Chemical Engineering, 2008. 32(8): p. 1956-1962.

13. Hu, D. and J.M. Yuan, Time-dependent sensitivity analysis of biological networks: coupled MAPK and PI3K signal transduction pathways. J Phys Chem A, 2006. 110(16): p. 5361-70.

14. Hua, F., et al., Integrated mechanistic and data-driven modelling for multivariate analysis of signalling pathways. J R Soc Interface, 2006. 3(9): p. 515--526.

See more of this Session: In Silico Systems Biology: Intracellular Signaling and Gene Regulation
See more of this Group/Topical: Topical A: Systems Biology

Lung Disease Diagnosis From Gene Expression Profiles

Shuyi Ma, Jaeyun Sung, Yuliang Wang and Nathan D. Price, Chemical and Biomolecular Engineering, University of Illinois at Urbana-Champaign, Urbana, IL

Abstract

Lung diseases include some of the most widespread and deadly conditions known to affect people in the US today. One of the main challenges in treating lung disease is the difficulty of diagnosis. Clinical diagnosis remains largely dependent upon symptomatic-based diagnoses; many cases can be either misdiagnosed or undiagnosed until disease has progressed to a more severe stage. Most studies aimed at finding molecular-based diagnostics have focused on one or two diseases at a time, yielding limited success. Instead, we searched for biomarkers reflective of the global health state of the lung by studying data taken from a broad range of lung diseases. We used gene expression microarray data from five different lung diseases—lung adenocarcinoma, lung squamous cell carcinoma, malignant pleural mesothelioma, chronic obstructive pulmonary disease, and asthma—as well as a non-diseased phenotype, to train a classification tree scheme based on the Top Scoring Pair (TSP) algorithm (Geman et al., Stat Appl Genet Mol Biol. 2004; 3: Article 19). The algorithm identified 27 gene pair classifiers that classify the three cancers explicitly, and several of the markers have been previously cited in literature as linked to these cancers. Ten-fold cross validation yielded a classification accuracy of approximately 88%. Thus, a TSP-based classification tree scheme accurately identifies lung diseases from the relative expression of a few number of diagnostic gene pairs.

Introduction

Lung diseases are some the one of the most pervasive and deadly diseases worldwide, causing over four million deaths each year[1]. In the US alone, asthma affects approximately 30 million people and is the most common chronic disease in children, [2]. Chronic obstructive pulmonary disease (COPD), which causes difficulty breathing, affects about 16 million people in and is the fourth leading cause of death[3]. Lung cancer is the most fatal cancer, leading to over 150,000 deaths per year [4].

Currently, a main challenge of treating lung disease is the difficulty of diagnosis. Most urgently, in the case of lung cancer, the tumor has often metastasized beyond the chest wall by the time of diagnosis. This late-stage diagnosis undermines the efficacy traditional treatments; only 17% of patients diagnosed with lung cancer survive more than five years[4].

A potentially valuable source of information to facilitate more effective diagnosis is the blood as it is comparatively non-invasive to sample and has been shown to contain organ-specific secreted gene products with concentrations that depend on disease state[5]. Despite the latent wealth of information found in blood, discovering viable blood markers for disease has so far proven daunting because current shotgun assay methods cannot adequately survey the entire set of blood proteins, which have concentrations that range approximately nine orders of magnitude[6].

We propose to harness computational algorithms to identify likely blood marker candidates based on gene expression microarray data. Many studies in search of molecular diagnostics have focused on characterizing one disease compared to a control[7]. This approach yields limited long term success because few biomarkers survive through clinical trials. We attempt a holistic approach by examining several lung associated phenotypes, both diseased and non diseased, to find context specific biomarkers that are indicative of specific lung health states rather than general markers of presence or absence of disease. Previous analyses of multiple diseases have examined expression profiles across hundreds of genes which, although informative, are too numerous to serve as biomarker candidates for blood screening. An alternative approach based on the top scoring pair (TSP) algorithm[8], which classifies based on relative expression of a handful of genes, is more desirable for identifying promising markers. Searching for likely blood markers with known identities facilitates targeted assays for more sensitive detection and quantification. We wish to leverage a classification scheme based on TSP analysis to identify potential blood markers that can diagnose multiple lung diseases with high accuracy. We will examine five different lung disease phenotypes—adenocarcinoma (ADC), squamous cell carcinoma (SCC), malignant pleural mesothelioma (PMT), asthma (AST), and COPD, as well as a non-diseased phenotype (NORM). Our analysis will be based on the gene expression microarray data of tissue samples collected from patients.

Methods

We assembled lung related expression data from the publicly available online database, Gene Expression Omnibus. To reduce inter-platform variation, we included only data collected from two Affymetrix microarray platforms: 1) HG-U133_Plus_2, and 2) HG-U133A, which share a large number of common probes (~ 20000) and have similar manufacturing procedures. We used these data to train the classification algorithm and to test its performance.

Our classification scheme is a machine learning algorithm based on TSP analysis developed by Geman et al.[8] and refined by Tan et al.[9] TSP analyzes ranked expression levels of samples for a binary phenotype classification. Geman's algorithm finds a single gene pair that displays maximal relative expression reversal between samples of the two phenotypes[8] (see Figure 1), as assessed by

$$TSPscore_{ij} = |p_{ij}(1) - p_{ij}(2)|,$$

where

$$p_{ij}(c) = P(Gene_i > Gene_j | c)$$

is the probability that $Gene_i$ has greater expression than $Gene_j$ in patients of phenotype c, relative frequency of occurrence in patients with known phenotype. Gene pairs with high $p_{ij}(c)$ for one phenotype and low $p_{ij}(c)$ for the other (i.e., high TSP scores) have the desired relative expression reversal that facilitates classification. Thus, the classifying gene pair with the highest TSP score is selected, and the classifying condition is: IF $Gene_i > Gene_j$ THEN phenotype 1, ELSE phenotype 2. The generalization of TSP, called k-TSP[9], selects a panel of n gene pair classifiers with the highest TSP scores, where n is a positive integer that can yield greatest classification sensitivity. A test sample that has an expression profile such that an integer $k \leq n$ of the gene pairs display the relative expression reversal indicative of phenotype 1 over phenotype 2 will be classified as phenotype 1. k is selected to maximize the specificity while maintaining sensitivity above a specified threshold.

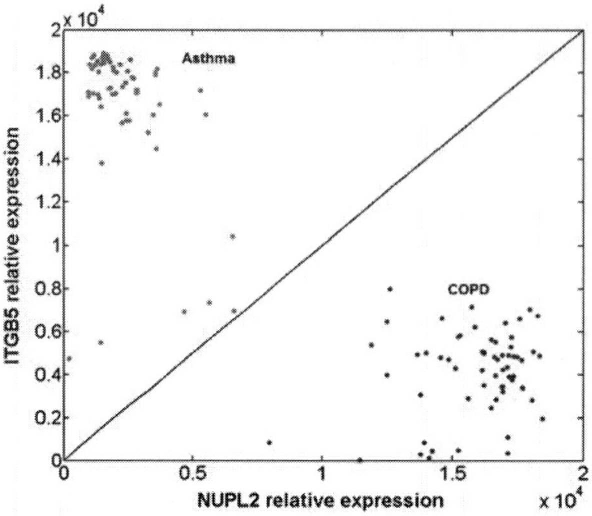

Figure 1: Standard TSP classification output applied to AST and COPD patients. The relative expressions of the classifying gene pair are plotted on the axes. In this case, the relative expression of the gene ITGB5 is always greater than NUPL2 in AST, and the opposite is observed in COPD.

TSP and k-TSP perform binary classification. To classify more phenotypes, we structure a series of binary TSP classifications into a hierarchical classification tree. The tree is constructed by successively grouping together phenotypes with the most similar expression profiles as quantified by TSP score. The complete tree has all considered phenotype classes in the topmost node. Branches that separate groups of phenotypes with the highest TSP score reside close to the top of the tree, and branches that separate groups with the lowest TSP score split at the bottom.

The classification tree informs two methods of sequential TSP comparisons. The first, node-based method focuses on the nodes of each tree. k-TSP is applied, yielding panels of gene pair classifiers that determine whether or not test samples are members of the phenotypes specified at each node. If a sample is affirmatively classified, then it passes through both branches spanning from the node and faces classification at both downstream nodes until it reaches nodes with only one phenotype. If a sample is not classified as one of the phenotypes at a node, then it is rejected from further classification at downstream nodes. This node-based classification can make multiple diagnoses because a sample could pass classification at multiple branches in the tree. Furthermore, rejection of a sample from being classified by the tree indicates that the sample does not appear to resemble any of the phenotypes in the tree.

The second, branch-based method compares groups on each left branch against groups on the corresponding right branch with TSP, yielding one gene pair classifier per branch. If a sample is classified as belonging to a phenotype in the left branch node, then it is passed only down the branches that are downstream of the left node. Otherwise, the sample is passed down for subsequent classification at the branches downstream of the right node. Because this method always provides a unique classification outcome, it is used to pick the best solution in the case where the node-based method yields multiple solutions

To test TSP tree algorithm performance, we executed a ten-fold cross validation by randomly excluding one tenth of the samples of each phenotype from the training data and by evaluating accuracy based on the classification of this excluded test set. The process is iterated ten times so that each excluded test set is disjoint.

Results

Table 1: This summarizes data collected from GEO. All data were gathered from two Affymetrix microarray platforms: 1) HG-U133_Plus_2, and 2) HG-U133A.

Disease	Label	Platforms	# Studies	# Samples	Sample Type
Adenocarcinoma	**ADC**	**1, 2**	**8**	**389**	resection
Squamous Cell Carcinoma	**SCC**	**1, 2**	**7**	**340**	resection
Pleural Mesothelioma	**PMT**	**1,2**	**2**	**49**	resection
Asthma	**AST**	**1**	**2**	**61**	airway epithelial brushing, bronchoalveolar lavage
Chronic Obstuctive Pulmonary Disease	**COPD**	**1, 2**	**4**	**63**	airway epithelial brushing, resection
Normal	**NORM**	**1,2**	**11**	**263**	airway epithelial brushing, resection

Table 1 describes the gene expression microarray data collected that are associated with each phenotype from the Gene Expression Omnibus (GEO), specifying the number of samples, the number of studies, the types of platforms, and the method of tissue extraction used to collect the samples. Each phenotype has data collected from two or more indepentent studies, with sample sizes ranging from 49 to 389. A potential source of variability originates from the heterogeneity of sample collection methods. Cancer tissue samples were collected from surgical resection, whereas some COPD samples were collected from airway epithelial endoscopic brushings. AST samples were collected by brushing or by bronchoalveolar lavage. NORM samples consist of tissue collected from sudden death autopsy subjects as well as controls from studies that supplied the diseased tissue samples. The surgical resection control tissues were collected from patients with lung cancer locations distal to the tumor.

We trained the TSP tree algorithm on these data, including both disease and non-disease phenotypes for consideration. In this paradigm, the top node does not have a classifier panel as all samples were assumed to belong to one of the six phenotypes. To maximize the chance of finding classifiers that can be measured in the blood, we restricted the number of genes considered to 1023 genes on the microarrays that code for peptides found in blood by the HUPO plasma proteome project[10]. We used a sensitivity threshold of 98% for finding optimal k. Figure 2 shows the TSP tree algorithm output consisting of the classification tree and gene pair classifier panels. The vote integer k at each node is equal to 1 unless otherwise specified. Notably, tree branching reflects phenotypic characteristics. For example, ADC and SCC, two forms of non-small cell lung cancer, are grouped together until the last branching. Similarly, AST and COPD, lung diseases associated with inflammation, are also grouped. Genes underlined in blue have citations that link classifiers to their associated phenotypes[11].

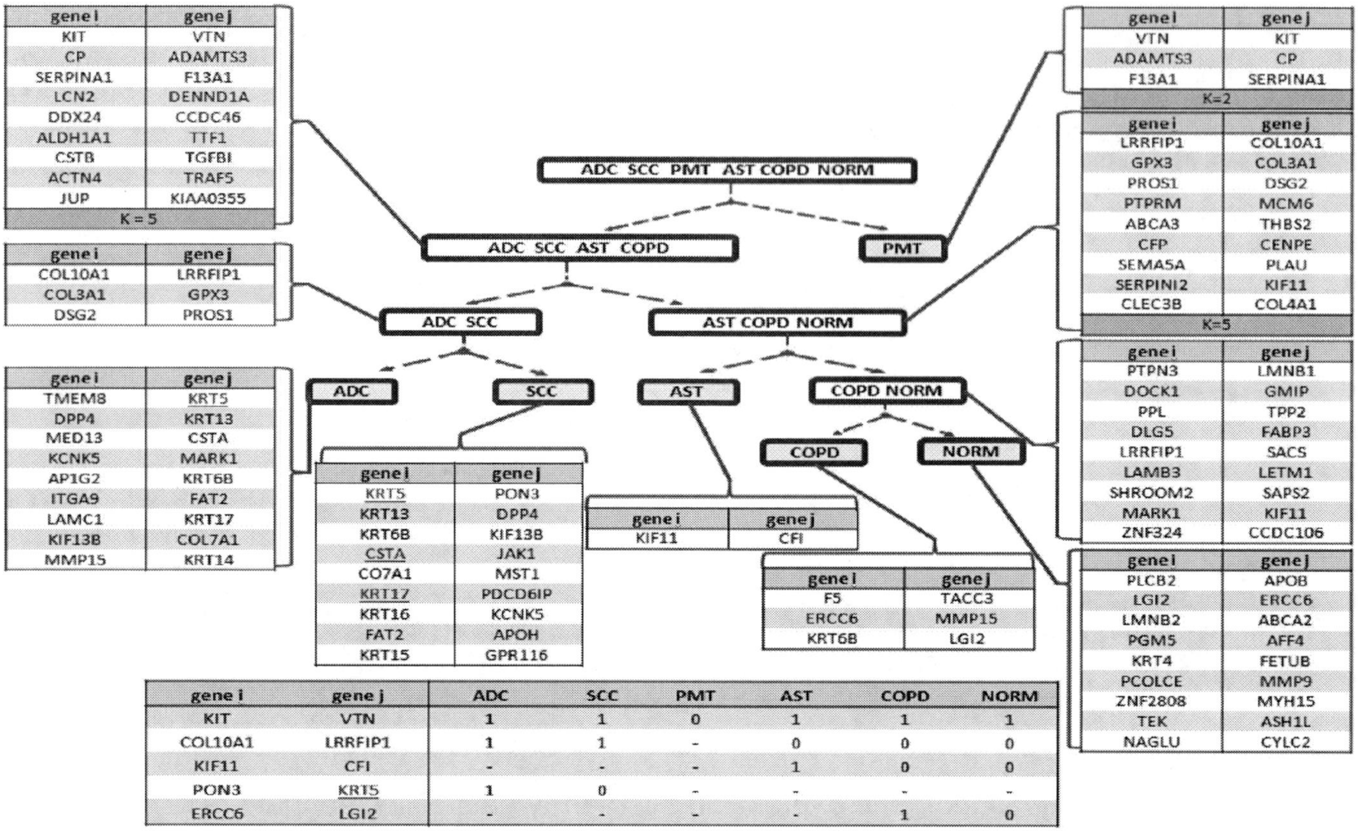

Figure 2: TSP classification tree and gene pair markers. The tables associated with each node specify the node-based classifiers. The table below shows the branch-based classifiers. The classifiers consist of 64 gene pairs of 100 unique genes.

Figure 3 compares ten-fold cross validation performance results of the TSP tree method against one-vs-one multiclass SVM. The TSP tree method has an average accuracy of 82%, which is comparable to the average performance of one-vs-one SVM: 83%. The TSP tree classifies lung disease expression data reasonably well, with 81% accuracy for SCC and COPD and at least 88% accuracy for the other three phenotypes. In contrast, classification accuracy of non-diseased samples is 46%, which while 2.76 fold better than chance, is still suboptimal.

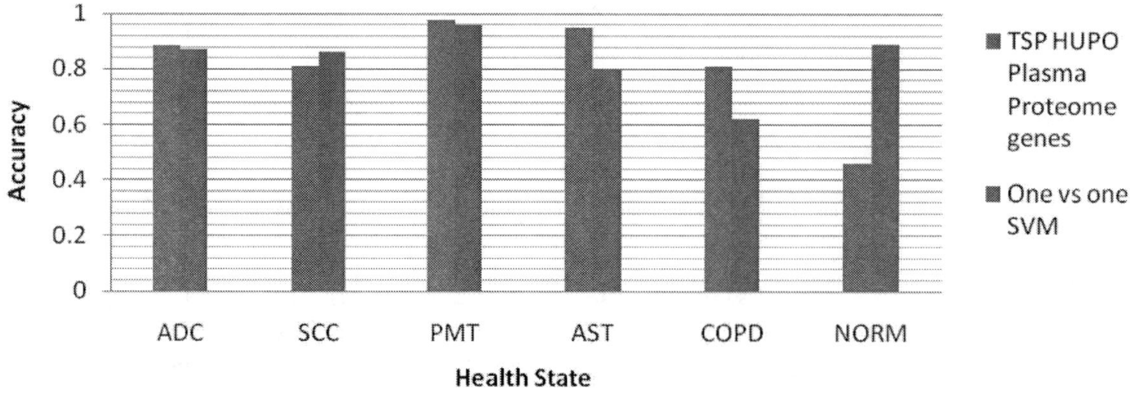

Figure 3: Ten-fold cross validation results of TSP tree algorithm compared to one-vs-one SVM. The TSP method performs better in classifying most of the diseases and worse in classifying non-disease.

The confusion matrix in **Error! Not a valid bookmark self-reference.** shows how samples of each phenotype are classified by the TSP tree algorithm. NORM samples have significant false positive rates (29%) for COPD. Ten-fold cross-validation analysis of COPD vs. NORM with TSP using the top 2500 differentially expressed genes and with SVM using all genes yielded 72% and 88% accuracy respectively, indicating that NORM samples are difficult to distinguish from COPD.

Table 2: The confusion matrix shows sample performance in ten-fold cross-validation. The rows represent the actual phenotype, and the columns represent the predicted phenotype. The red values highlight the substantial error sources of NORM misclassification.

		Predicted Class							
		ADC	SCC	PMT	AST	COPD	NORM	REJECT	Total
Actual Class	ADC	0.88	0.08	0.01	0	0.02	0.01	0.00	389
	SCC	0.15	0.81	0.02	0.003	0.01	0.01	0	340
	PMT	0	0	0.98	0	0	0	0.02	49
	AST	0.02	0	0.02	0.95	0.02	0	0	61
	COPD	0.05	0.11	0	0	0.81	0.02	0.02	63
	NORM	0.03	0.02	0.03	0.16	0.29	0.46	0.01	263
	Total	407	317	64	102	142	127	6	

Accounting for the difficulty in separating NORM from COPD and AST to a lesser extent, we constructed a second TSP tree that explicitly classifies only the cancer phenotypes (Figure 4). Unlike the tree containing all phenotypes, classification is performed at the top node, which distinguishes whether a sample is one of the lung cancers considered or a non-cancer. There is substantial overlap between the gene pair classifiers in this tree and those in the large tree, including the genes that are cited to be related to their respective phenotypes[11].

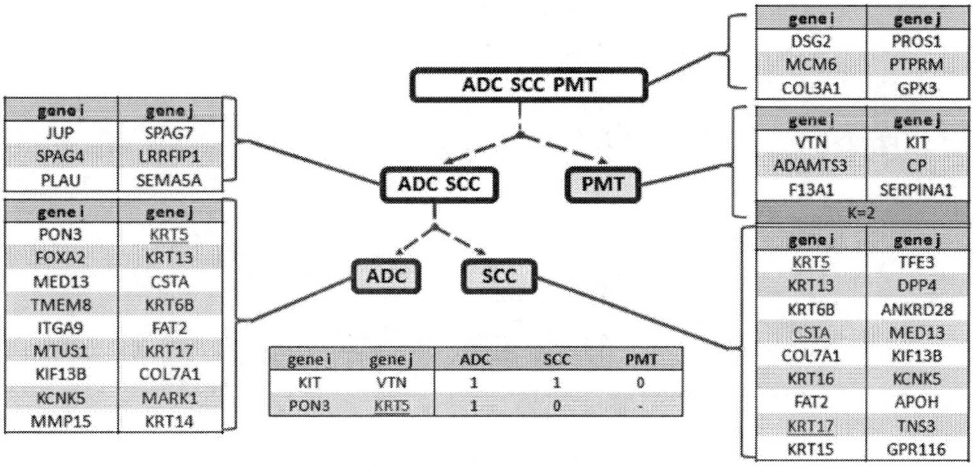

Figure 4: TSP tree classification scheme for lung cancers. The classifiers consist of 27 gene pairs of 44 unique genes.

Figure 5 compares ten-fold cross-validation performance of the TSP cancer tree to one-vs-one SVM classification. Performance on non-cancer phenotypes is measured as the fraction of samples correctly rejected from further classification in the cancer tree. The TSP tree method has an average accuracy of 88%, which is higher than the average performance of one-vs-one SVM: 83%, and it performs with higher accuracy than SVM in three of the five phenotypes. The cancer tree rejects non-cancerous phenotypes with accuracies of 84-92%. The poorest overall performer, SCC has an accuracy of 81% with the main source of error deriving from misclassification as ADC, possibly due to a shared non-small cell lung cancer signature. COPD has the greatest non-cancerous false rejection rate and is mistaken for both ADC and SCC with 8% frequency for each phenotype, possibly due to a shared smoking signature.

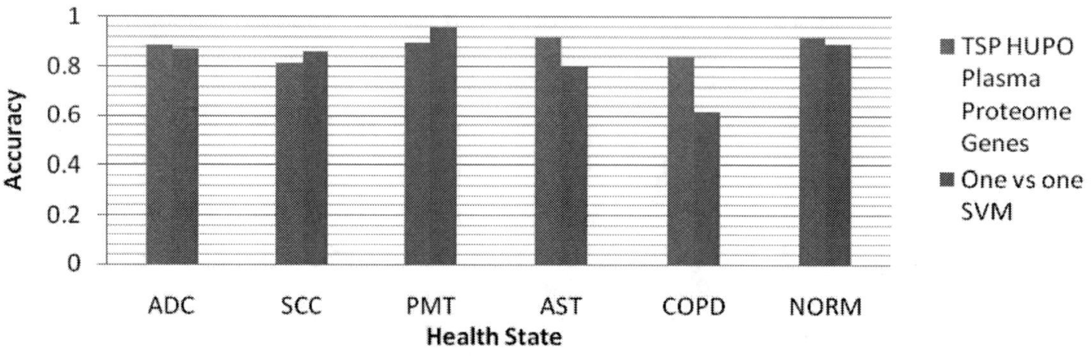

Figure 5: Ten-fold cross validation results of TSP cancer tree and SVM classification.

To examine the correlation of gene expression level with protein level of the TSP cancer tree classifiers, we examined immunohistochemistry stains for ADC and SCC from the Human Protein Atlas[12] (Figure 6). Visual inspection indicates that many marker pairs display the same relative expression profiles as predicted by TSP. This suggests that these markers might have similar predictive power when measured on the protein level.

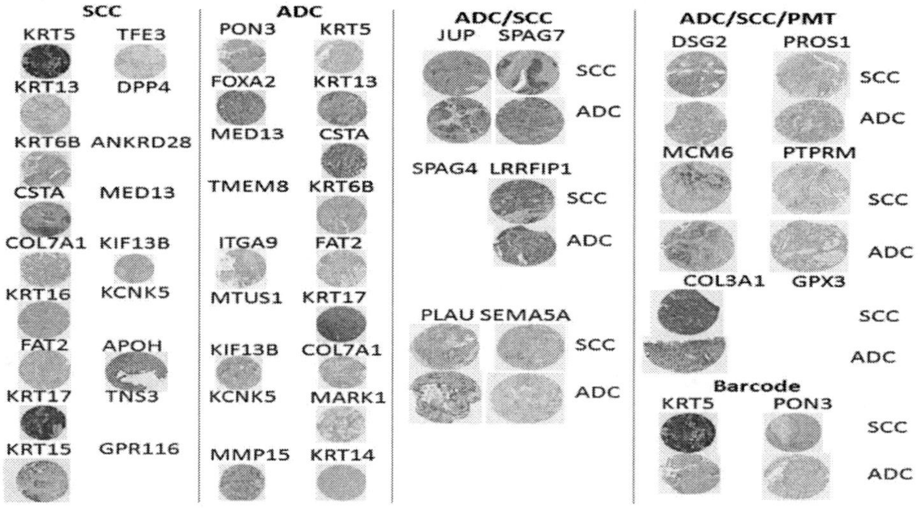

Figure 6: Immunohistochemistry stains of ADC and SCC classifiers. Brown indicates presence of protein and blue shows cell nuclei. Markers in each left column have greater expression than markers in the right columns in the TSP cancer tree method. Blanks indicate that no stains exist in the database.

Discussion

We have applied a TSP classification tree algorithm to diagnose lung diseases from gene expression microarray data. Our first tree scheme explicitly diagnoses all phenotypes. This tree classifies disease phenotypes with high accuracy but falsely classifies non-diseased samples as being diseased more than half of the time. The misclassification occurred because many NORM samples molecularly resembled COPD and AST samples. This might be because a substantial number of NORM samples were non-cancerous tissue taken from lung cancer patients. These subjects might have had undocumented COPD or AST. Also, a significant number of samples were controls from studies examining AST and COPD. COPD and AST had comparatively few numbers of samples and independent studies undertaken, which potentially allows for a lab-specific signature to dominate the expression profiles. Given the difficulty separating NORM from COPD and AST, we chose to focus on the lung cancers.

Our second tree scheme explicitly diagnoses only the cancer phenotypes and rejects non-cancers from classification. This cancer tree classifies the phenotypes examined with comparable or better accuracy than one-vs-one SVM. Furthermore, immunohistochemistry stains showed that the gene products of many classifiers display similar relative concentrations to those predicted by TSP, which substantiates the possibility that these classifiers have diagnostic protein concentration profiles in the blood.

Interestingly, several classifiers isolated by the algorithm can be linked to their disease phenotypes in literature. This suggests that there might be an underlying biological reason why the gene pairs are diagnostic. We plan to examine further the potential biological significance of classifiers, especially in the context of probing the underlying gene and protein networks that are perturbed by disease.

We have taken the first step toward finding diagnostic blood markers for lung disease. Guided by the lists of gene pairs inferred from the TSP tree algorithm, we can target sensitive assays to quantify protein levels of promising classifiers in the blood. If a set of blood markers can demonstrate predictive reproducibility, these can be harnessed as a platform to aid effective diagnosis and presymptomatic screening of disease. Moreover, elucidating the potential contributions of the markers to disease-inducing perturbations might bring greater insight into underlying mechanisms of the diseases and inform a closer monitoring of disease states. Thus, markers potentially have diagnostic utility to aid effective early screening of patients and to assess the state of diseases through its progression and through treatments.

References

1 The top 10 causes of death, Available at http://www.who.int/mediacentre/factsheets/fs310/en/index.html, (2008).

2 Ettinger, D. S., in *Cecil Medicine*, edited by L. A. Goldman, D. A. et al. (Saunders, 2008).

3 COPD Statistical Information, Available at http://www.copd-international.com/library/statistics.htm, (2004).

4 Lung Cancer, Available at http://www.emedicinehealth.com/lung_cancer/article_em.htm, (2010).

5 Leidinger, P. et al., *Respir Res* **11**, 18.

6 Adkins, J. N. et al., *Mol Cell Proteomics* **1** (12), 947 (2002).

7 Maciel, C. M. et al., *J Exp Ther Oncol* **5** (1), 31 (2005); Molina, R. et al., *Tumour Biol* **24** (4), 209 (2003).

8 Geman, D., d'Avignon, C., Naiman, D. Q., and Winslow, R. L., *Stat Appl Genet Mol Biol* **3**, Article19 (2004).

9 Tan, A. C. et al., *Bioinformatics* **21** (20), 3896 (2005).

10 Omenn, G. S. et al., *Proteomics* **5** (13), 3226 (2005).

11 Wei, K. J. et al., *Oral Dis* **15** (1), 111 (2009); Medina, P. P. et al., *Hum Mol Genet* **18** (7), 1343 (2009); Kargi, A., Gurel, D., and Tuna, B., *Appl Immunohistochem Mol Morphol* **15** (4), 415 (2007).

12 Berglund, L. et al., *Mol Cell Proteomics* **7** (10), 2019 (2008).

(597b) F-Test in k-Fold Cross Validation and Its Application to the Discovery of Biological Networks

Mano R. Maurya, Department of Bioengineering, University of California San Diego, La Jolla, CA, Shakti Gupta, Department of Bioengineering, University of California, San Diego, CA and Shankar Subramaniam, San Diego Supercomputer Center, Department of Bioengineering, Department of Chemistry & Biochemistry, University of California San Diego, La Jolla, CA

There has been considerable emphasis in the recent years on applying systems approaches to decipher and reconstruct cellular networks using high-throughput data. To avoid over-fitting the data and to ensure that the resulting model has good predictive power, cross-validation is often used in data-driven input-response (or input/output (I/O)) modeling. F-test is commonly used to compare the fit-errors (usually, sum of squared prediction errors (SSE)) of the model on the training and test sets. If sufficiently large dataset is available then the data can be divided into non-overlapping training and test sets and F-test can be applied subject to the assumption of the normality of the experimental data points and hence that of the prediction errors. However, when large datasets are not available owing to the cost of conducting experiments, as is often the case for biological systems, k-fold cross validation (CV) is used. In k-fold CV, the entire dataset is randomly divided into k groups. The model is developed using (k - 1) groups as training set and remaining one set is used as the test set. This process is repeated until all k groups are used as a test set once. The mean of the SSE for the test set is compared with the mean of the SSE for the training set through F-test. In this case, 1/k fraction of the samples in any training set is exactly the same as 1/k fraction in other (k-1) training sets. Hence, the computation of the degree of freedom (DOF) for the average SSE for the training sets is not straight-forward. To the best of our knowledge, in most existing work on k-fold CV, the comparison between the average SSE for the training and test sets is carried out qualitatively in an ad-hoc fashion. In this work, we have developed a rigorous procedure to compute the DOFs for robust F-test in k-fold cross-validation.

We have used this approach of k-fold CV to a partial-least squares (PLS)-based method for identifying the interactions between different signaling proteins using phosphoprotein data in mouse macrophage RAW 264.7 cells provided by the Alliance for Cellular Signaling (AfCS). A value of k = 10 was used. In the PLS-based modeling scheme used here, only one output is used at a time (1), which is different from the traditional way of applying PLS technique on I/O data. Once the I/O model is deemed robust based on the F-test, significant interactions are selected through t-test (1, 2) and are used to reconstruct the phosphoprotein signaling network. Important signaling events such as activation of glycogen synthase kinase 3 by protein kinase B (Akt) are captured by our reconstructed network. Novel links as well as testable hypotheses are also generated by our analysis approach. We will also show the application of the approach to least-square regression and principal component regression-based techniques for modeling I/O data.

Reference

1. Gupta, S., M. R. Maurya, and S. Subramaniam. 2010. Identification of crosstalk between phosphoprotein signaling pathways in RAW 264.7 macrophage cells. PLoS Comput Biol. 6:e1000654.

2. Pradervand, S., M. R. Maurya, and S. Subramaniam. 2006. Identification of signaling components required for the prediction of cytokine release in RAW 264.7 macrophages. Genome Biol. 7:R11.

See more of this Session: In Silico Systems Biology: Intracellular Signaling and Gene Regulation II
See more of this Group/Topical: Topical A: Systems Biology

(597c) Sensitivity Analysis of a Kinetic Model of Interactions Among Ca2+, Calmodulin and CaMKII

Tamara L. Kinzer-Ursem[1], Shirley L. Pepke[2], Stefan Mihalas[3] and Mary B. Kennedy[1], (1)Biology, California Institute of Technology, Pasadena, CA, (2)Center for Advanced Computing Research, California Institute of Technology, Pasadena, CA, (3)Mind/Brain Institute, Johns Hopkins University, Baltimore, MD

Changes in the strength of synaptic connections in the brain underlie our ability to form memories and to learn. One type of experimentally induced change in synaptic strength, long term potentiation (LTP), is dependent upon the activation of Ca2+/calmodulin dependent protein kinase II (CaMKII). CaMKII is a serine/threonine protein kinase that constitutes 1-2% of all brain protein by weight. It is activated by binding of the Ca2+/calmodulin (Ca2+/CaM), which binds up to four Ca2+ ions upon Ca2+ flux through the NMDA receptor in postsynaptic dendritic spines. CaM removes the inhibitory domain of the kinase from the catalytic site and allows CaMKII to autophosphorylate itself and phosphorylate its substrates. CaMKII is activated within milliseconds of Ca2+ influx and can remain active for tens of minutes afterward. It is an early and essential molecular component of the complex signal transduction processes that underlie LTP.

Here we present a thermodynamically complete model of activation of monomeric subunits of CaMKII (mCaMKII) by Ca2+/CaM that includes binding of Ca2+ to free CaM and to CaM bound to individual mCaMKII subunits. The complete model of CaMKII activation accounts for the different kinetics of Ca2 + binding to the amino (N) and carboxyl (C) termini of CaM, and for the thermodynamic stabilization of Ca2+ binding when CaM is bound to a target protein, in this case mCaMKII and the autophosphorylation of mCaMKII for a total of 154 protein states. The values of the kinetic rate constants describing the interactions are well constrained by previous experimental studies; however, a few have not been measured directly. In these cases, we fit existing experimental data and used the principle of microscopic reversibility to derive reasonable ranges of values. We used the model to make predictions about the time course of autophosphorylation of CaMKII under conditions that are believed to exist in synaptic spines and under commonly used experimental "test tube" concentrations of Ca2+, CaM, and mCaMKII.

We performed global variation and sensitivity analyses to determine which parameters most affect the levels of autophosphorylation under various conditions. We have used these analyses to infer the kinetic pathways through which autophosphorylation of CaMKII is likely to proceed. Under conditions that are believed to prevail in a spine during activation of NMDA receptors, both the level of Ca2+ and the amount of available CaM limit the rate of autophosphorylation of mCaMKII. This means that competition for Ca2+/CaM in an activated spine is an important determinant of synaptic plasticity because autophosphorylation of CaMKII is critical for induction of LTP. This work shows that kinetic analyses can add to our understanding of the complex mechanisms controlling autophosphorylation of

CaMKII in spines; and, more generally, should help to clarify other Ca2+/CaM signaling events. Additionally, the model presented here is a first step in a larger project to build kinetic simulations of activation of the CaMKII enzyme in the context of physiologically realistic models of Ca2+ fluctuations in postsynaptic spines. To this end, we have used sensitivity analysis to identify parameters whose refinement by direct measurement will be most important for the accuracy of predictions from our current and future models.

See more of this Session: In Silico Systems Biology: Intracellular Signaling and Gene Regulation II
See more of this Group/Topical: Topical A: Systems Biology

(601a) Is It Indole Language or Food? Phase Plane Analysis of Indole as Quorum Sensing Signal in E. Coli

Daniela Florez and **Andrés Fernando Gonzalez Barrios**, Ingenieria Química, Universidad de los Andes, Grupo de Diseño de Productos y Procesos (GDPP), Bogota, Colombia

Indole is a stationary phase signal secreted by Escherichia coli and has been shown to have two important roles: quorum sensing signal that reduces biofilm formation and metabolic reporter that regulates several genes expression (e.g., astD, tnaB, gabT). Nevertheless, the main function that carries out indole in E. coli has not been clearly elucidated. Here we evaluate its role through computational and experimental techniques utilizing Flux-balance analysis (FBA) model to determine metabolites flux distribution relying on metabolic reactions stoichiometry, limited substrates availability and optimality principles. We found that synthesis and exchange fluxes which maximize the biomass production were 24.6 and 2.64 mM/h g cell, respectively. The shadow prices vector was calculated to study the biomass objective function sensibility to finally obtain a phenotype phase plane. In order to validate the plane obtained we evaluated biofilm formation in 96-well plates and cellular growth at different exchange and synthesis flux of indole by increasing extracellular indole concentration or modifying E. coli indole flux synthesis with E. coli K-12 pCA24N tnaA+. The indole synthesis process displays a stronger impact than exchange process on metabolic state of E. coli. So we conclude that indole in E. coli displays a role as metabolic state probe rather than quorum sensing signal.

See more of this Session: Modeling Approaches to Examine Fundamental Issues in Life Sciences - I
See more of this Group/Topical: Food, Pharmaceutical & Bioengineering Division

(601b) An Extended Data Mining Method for Identifying Differentially Expressed Assay-Specific Signatures in Functional Genomic Studies

Derrick Rollins and AiLing Teh, Iowa State University, Ames, IA

Microarray data sets provide relative expression levels for thousands of genes for a small number, in comparison, of different experimental conditions called assays. Data mining techniques are used to extract specific information of genes as they relate to the assays. The multivariate statistical technique of principal component analysis (PCA) has proven useful in providing effective data mining methods. This talk extends the PCA approach of Rollins et al. (2006) to the development of ranking genes of microarray data sets that express most differently between two biologically different groupings of assays. This method is evaluated on real and simulated data and compared to a current approach on the basis of false discovery rate (FDR) and statistical power (SP) which is the ability to correctly identify important genes. This work developed and evaluated two new test statistics based on PCA and compared them to a popular method that is not PCA based. Both test statistics were found to be effective as evaluated in three case studies: (i) exposing E. coli cells to two different ethanol levels; (ii) application of myostatin to two groups of mice; and (iii) a simulated data study derived from the properties of (ii). The proposed method (PM) effectively identified critical genes in these studies based on comparison with the current method (CM). The simulation study supports higher identification accuracy for PM over CM for both proposed test statistics when the gene variance is constant and for one of the test statistics when the gene variance is non-constant. PM compares quite favorably to CM in terms of lower FDR and much higher SP. Thus, PM can be quite effective in producing accurate signatures from large microarray data sets for differential expression between assays groups identified in a preliminary step of the PCA procedure and is, therefore, recommended for use in these applications. It is well known that living organisms have complicated gene structures. However, while major advancements have been made in recent years, understanding of the biological functions of each individual gene is still quite limited. Active research is strongly focused on understanding the behavior of genes and as well as the highly complex metabolism and regulatory network inside living cells. This effort falls under a molecular biological field called functional genomics (FG). There are at least three areas in which experimental techniques are widely applied in FG: transcriptomics, proteomics, and metabolomics. A combination of leading scientific techniques as well as powerful mathematical and statistical tools for data analysis makes the task of identifying important transcriptome, proteome, and metabolome corresponding to a biological effect promising. Typical studies in these areas involve the identification of possible behavior and responses of species under various genetic backgrounds as well as environmental factors (i.e. assay). There are different high technology techniques applied in FG field to advance understanding of the transcriptional genetic response of many organisms in various environmental perturbations. One of the techniques that have been adopted in this field is a multiplex technology called DNA microarray. A new technique that is becoming popular and will probably displace array-based measurement in FG is next-generation sequencing (RNAseq). These techniques have the ability to generate data sets that consist of expression levels of thousands of genes, providing a wealth of information that is hidden by high noise levels, low signal levels, and a relatively small number of experimental units to the number of genes studied. More specifically, since the data set containing the gene expression measurements consists of a lot more genes than assays, analytical techniques are needed to provide accurate gene identification under a large number of gene candidates that is much greater than the number of experimental runs. To achieve this objective, traditional statistical methods, such as principal component analysis (PCA), the focus of this article, are being retrofitted to provide effective statistical inference in this challenging context of microarray data analysis. Other methods used in this field included linear model analysis, Bayesian method and neural network analysis (NCA). Thus, statistics is playing a critical role through the development of methodologies that give high statistical power (SP) (i.e., accurate identification), and low false discovery rate (FDR) (i.e. low misidentification). To this end, this talk introduces two new PCA based statistics for determining gene rank for differential expression between two PCA identified assay groups. This work extends the technique introduced by Rollins et al. (2006) that determines gene rank for a single PCA identified assay group. Thus, the proposed method (PM) in this work is aimed at finding the genes with high expression levels in one group and low expression levels in the other group. The PM uses PCA to first establish the existence of the assay groupings of interest. Then using the results that established the grouping, the differential contribution for each gene is determined using a statistic based on eigenvalues. This talk discusses and evaluates two statistics. The first one is the group averaged difference of eigenvalue linear combinations that we call Tdiff. The second one divides Tdiff by its estimated pooled standard deviation that we call Tscaled. The genes are ranked based on the largest absolute value of these statistics. The PM is evaluated against the ranking determined by the well known Student's t-statistic that we call Tpooled in this work. We will refer to Tpooled as the current method (CM) which is actually a subclass of the PM that weighs each assay equally in each group. Note that for the CM the assay members in each group is not established based on the data but by á priori considerations. In contrast, for the PM the data drives the assay weight as well as group assignment of the assays. The CM and PM are applied in the following three case studies to compare their effectiveness (i.e., power) in identifying assay-specific signature: (i) exposure of E. coli cells to two different levels of ethanol concentration; (ii) the use of myostatin as inhibitor of skeletal muscle growth for five 5-weeks-old myostatin and non-treated mice; and (iii) a simulation study based on statistical properties of the second case study. This talk will give a brief review of PCA and connect it to our application in FG's data analysis. Next we will derive and present the test statistics of the CM and PM. These test statistics will be evaluated and compared in three studies.

See more of this Session: Modeling Approaches to Examine Fundamental Issues in Life Sciences - I
See more of this Group/Topical: Food, Pharmaceutical & Bioengineering Division

(601c) Designing Cryoprotectant Addition and Removal Protocols for Vitrification of Tissue-Engineered Constructs Using a Mathematical Model

Alison Lawson, Chemical & Biomolecular Engineering, Georgia Institute of Technology, Atlanta, GA, Indra N. Mukherjee, Merck, Philadelphia, PA and Athanassios Sambanis, School of Chemical & Biomolecular Engineering, Georgia Institute of Technology, Atlanta, GA

In the field of Tissue Engineering, many different technologies are being investigated with the goal of bringing replacement tissues from the benchtop to the clinic. One of the most critical technologies, cryopreservation, is often overlooked. Cryopreservation will be necessary for the storage of any natural or artificial tissue. Any tissue headed for clinical use will require sterility and quality control testing as well as transport. It will be necessary to maintain both the cells and extracellular matrix during this time. The most common type of cryopreservation, conventional freezing, allows for ice formation, which may be damaging to the extracellular matrix. More recently, vitrification has been used as a cryopreservation method that completely avoids ice formation. Vitrification utilizes high concentrations of cryoprotectants (CPAs) and rapid cooling to achieve a vitreous, or glassy, state. Because of the high concentrations necessary to achieve reliable vitrification, the design of addition and removal protocols is critical for the success of this process. Cytotoxicity and cellular osmotic excursions must be minimized so as to minimize cell death, but it is also important to ensure that a vitrifiable concentration of CPAs is present throughout the construct and within the cells in the construct. Currently, most protocols are based primarily on the prior experience of the researcher who makes a "best guess" at a protocol that will maintain cellular osmotic excursions within a tolerable limit. This trial-and-error approach may allow coarse adjustment of protocols but does not allow for finer tuning that may result in better performing addition and removal procedures. A mathematical model would allow for fine tuning, as well as eliminate the cumbersome and time-consuming trial-and-error experiments. A mathematical model also allows much more freedom to investigate than experiment. More complex situations such as ramp or step changes in temperature or other conditions may be difficult to accomplish experimentally but may also prove to have significant benefit. Therefore, using a model should give insight into how beneficial complex conditions may be and whether they should be pursued experimentally. Previously, a model based on mass transport and cell permeability with experimentally determined parameters has been used to design addition and removal protocols (Mukherjee 2008). Here, we present an expanded version of this model, which additionally incorporates heat transfer and CPA cytotoxicity—both critical processes—in order to give a more complete picture of the effects of temperature and time on the addition and removal protocol. The hypothesis is that a model including non-isothermal conditions and CPA-mediated cytotoxicity will improve the design of addition and removal procedures by minimizing cell death. The basic structure of a model comprised of mass transport through the matrix and cell membrane, and heat transport through the tissue, can be used for a variety of natural and artificial tissues with the appropriate parameter values in each case. In this work, the model was applied to two tissue systems, which represent extreme situations in the CPA effective diffusivity through the matrix. The first system considered was an engineered tissue composed of murine insulinomas microencapsulated in hydrogel beads. In this, the effective diffusivity through the microcapsules was high, 80% of the bulk solution diffusivity, and the cells were suspended in the matrix. Several model parameters were experimentally determined and incorporated into the model. Cell permeability parameters were determined by measuring cell volume as a function of time after the abrupt addition of dimethyl sulfoxide (DMSO) or propanediol (PD) in a cell suspension. Effective diffusivities through the matrix were determined by adding a CPA into a suspension of capsules and measuring the decline in bulk concentration over time. Cytotoxicity studies were done with DMSO, PD and vitrification cocktails in order to determine the relationship between cell viability and concentration, temperature and time. The second system considered was natural articular cartilage which consists of chondrocytes in a dense extracellular matrix with an effective diffusivity only 30% of the bulk solution. Parameter values for this system were obtained from the literature. Currently, mathematical simulations have focused on situations that are easy to realize experimentally, as well as more complex non-isothermal conditions. Ramp changes are of particular interest as temperature affects both mass transfer and cytotoxicity and may change the overall performance of the addition and removal protocol. Simulating different situations also allows for determination of the sensitivity of the protocol. For example, the effects of over-exposure of CPAs differ based on the temperature and the CPA itself. Protocols using the less cytotoxic CPAs should be less sensitive to temperature changes. Unfortunately, the less cytotoxic CPAs also generally require higher concentrations to achieve a glassy state. Therefore, a balance must be found that minimizes the cytotoxicity while still ensuring vitrification. Use of the model to investigate these allows determination of the most critical factors for each system. Vitrification itself depends upon sufficient mass transfer and heating and cooling rates. The mathematical model describing the addition and removal of CPAs allows for design of protocols that achieve a sufficient concentration of CPAs throughout the construct and cells. This makes simulation of the addition and removal protocols critical to the successful vitrification of a tissue-engineered construct. The mathematical model was used to describe the performance of different addition and removal protocols for two vastly different constructs at isothermal and non-isothermal conditions. The results from these simulations allow insight into the future design and implementation of different vitrification solutions for use in the preservation of natural and engineered tissues.

See more of this Session: Modeling Approaches to Examine Fundamental Issues in Life Sciences - I
See more of this Group/Topical: Food, Pharmaceutical & Bioengineering Division

(601d) Development of a Mathematical Model to Describe the Transport of Monocyte Chemoattractant Protein-1 through a Three-Dimensional Collagen Matrix

Krisada Leemasawatdigul, Neda Ghousifam and Heather Fahlenkamp, School of Chemical Engineering, Oklahoma State University, Stillwater, OK

Cellular adhesion molecules and inflammatory cytokines are critical participants in the vascular dysfunction and tissue injury associated with a wide variety of inflammatory and cardiovascular diseases, including atherosclerosis. Atherosclerosis is an inflammatory disease in which its initiation involves with the subendothelial accumulation of lipid substances, and the transendothelial migration of monocytes in an artery. Eventually, the monocytes differentiate into cholesterol-engorged macrophages called "foam cells" that result in the atherosclerotic plaque formation. Having a better understanding of the underlying mechanisms and identifying potential pro-atherogenic markers will help to target these adhesion molecules or cytokines with highly specific therapeutic strategies. Monocyte chemoattractant protein-1 (MCP-1 or CCL2) is a chemokine that is expressed highly in atherosclerotic lesions. The key function of MCP-1 in the formation and progression of the disease is to recruit monocytes to a lesion area. Results of many studies have shown that monocyte transmigration is directed by the soluble gradient of MCP-1. In this study, a 3D in vitro human vascular tissue model, consisting of a collagen matrix, to mimic the subendothelial extracellular matrix (ECM), was introduced to investigate the formation of MCP-1 gradients in the ECM. The 3D tissue model provides the added dimension that is important for the creation of concentration gradients, along with cellular movement and interactions created by such gradients. The main objective of the study is to develop a mathematical model to determine the MCP-1 concentration in the collagen matrix of the tissue model. Both the unsteady-state mass transfer and the kinetics of MCP-1 in the collagen matrix were included in the mathematical model. To characterize the kinetic part of the model, experiments were performed to examine the stability of MCP-1 and the binding reaction between MCP-1 and the collagen matrix. Experimental results show that MCP-1 is stable at standard cell culture conditions (37°C, humidified atmosphere of 5% CO_2 and 95% air) for at least 24 hours, and it does bind with the collagen matrix. These findings were taken into account in the development of the mathematical model. Results from the model suggest that not only the soluble gradient of MCP-1 is formed in the collagen matrix, but a static gradient (the gradient of collagen-bound MCP-1) appears in the matrix as well. The static gradient may be another factor that controls the migration of monocytes into the lesion area.

See more of this Session: Modeling Approaches to Examine Fundamental Issues in Life Sciences - I
See more of this Group/Topical: Food, Pharmaceutical & Bioengineering Division

(601e) Physiological, Proteomic, and in Silico Cost-Benefit Analysis of Resource-Limited Growth

Ross P. Carlson[1], Reed Taffs[2] and James Folsom[1], (1)Chemical and Biological Engineering, Montana State University, Bozeman, MT, (2)Chemical Engineering, Montana State University, Bozeman, MT

Evolutionary selection has produced fit microbes capable of adapting their metabolic functioning to cope with competitive environments while utilizing limited resource pools. Physiological and proteomic characteristics of Escherichia coli cultures grown in either nitrogen- or iron-limited chemostat conditions were predicted and interpreted using an in silico cost-benefit analysis. The in silico analysis used ecological theory that examines resource allocation strategies that maximize fitness. All potential strategies for investing nitrogen and iron into the metabolic machinery of E. coli were enumerated by decomposing the metabolic network into a complete listing of non-divisible, mathematically-defined biochemical pathways. The enzymatic resource investment requirements for each distinct pathway were examined in concert with the pathway's efficiency at converting substrate into biomass. Allocating limiting resources to perform one function well typically comes at the cost of performing other metabolic functions well. The analysis identified competitive molecular-level tradeoffs between pathway resource requirements and metabolic efficiency that are likely important to ecological versatility. Experimental chemostat data illustrated that under nitrogen- and iron-limited conditions E. coli regulates its metabolism to invest the limiting resource competitively at the cost of optimal biomass yields on electron donor. The study highlights fundamental design considerations for effective control and engineering of microbes for bioprocess and biomedical applications.

See more of this Session: Modeling Approaches to Examine Fundamental Issues in Life Sciences - I
See more of this Group/Topical: Food, Pharmaceutical & Bioengineering Division

(601f) Computational Analysis of the Compartmentalization of Phosphatase-Mediated Regulation of the Epidermal Growth Factor Receptor

Calixte S. Monast and Matthew J. Lazzara, Chemical and Biomolecular Engineering, University of Pennsylvania, Philadelphia, PA

Cell signaling via the epidermal growth factor receptor (EGFR) is broadly involved in development, normal physiology, and pathology. Phosphorylated EGFR (pEGFR) levels, which dictate receptor signaling strength and persistence, are the net result of two opposing processes, auto-phosphorylation of cytoplasmic tyrosines by the receptor kinase and dephosphorylation of those tyrosines by protein tyrosine phosphatases. Several studies have concluded that pEGFR dynamics are regulated by phosphatases mainly in the cell interior, subsequent to receptor endocytosis. Here, we challenge this view through the development of a novel mechanistic model of EGFR phosphorylation. Our model is based on a set of coupled ordinary differential equations that describe receptor-level processes including ligand binding, competitive binding between ATP and kinase inhibitors, receptor dimerization, tyrosine phosphorylation, receptor dephosphorylation at the cell membrane and in the cell interior, and receptor trafficking. The list of model reactions was chosen to minimize the number of model parameters and species, while still incorporating key molecular rate processes. We found that model results were in general unchanged by expanding the model scope to include molecular species that might be predicted to form but which were excluded from the model. The few model parameters which were not taken from the literature were estimated by performing a multi-parameter fit of the model to several different types of experimental data taken from the literature using a simulated annealing algorithm. Whereas virtually all previous models of pEGFR dynamics involve fits to measurements of relative changes in pEGFR levels in response to ligand alone, we simultaneously fit our model to data of pEGFR dynamics in response to ligand, phosphatase inhibitors, or EGFR kinase inhibitors. In addition to these dynamic data, the model was also fit to measurements of basal pEGFR levels and actual percentages of total EGFR phosphorylated for particular conditions. Finally, the fit also involved an approach to recapitulate measurements of EGFR endocytosis rate constants which is more consistent with the definition of the experimentally measured rate constant than has been implemented in previous models. Use of this broad collection of different types of data for pEGFR levels represents a significant departure from previous modeling approaches. Comparison of our fitted parameters to those from previous studies revealed several noteworthy results. First, the difference between our fitted constants for receptor dimerization kinetics in the presence or absence of ligand was significantly greater than that reported in previous modeling studies, a difference which results mainly from our incorporation of kinetic data for receptor dephosphorylation in the fit. A more surprising result was that the fitted rate constants for phosphatase kinetics suggested that receptor dephosphorylation may actually occur more rapidly at the cell surface than in the cell interior. Based on these fitted parameters, the model further predicted that such high phosphatase activity at the cell surface may have a profound effect on EGFR endocytosis kinetics. Specifically, by reducing the concentration of EGFR phosphotyrosines at the cell surface, which are a primary driving force for receptor internalization, phosphatase activity at the cell surface was predicted to slow the net rate of EGFR endocytosis. This prediction is qualitatively consistent with a recent report of phosphatase-mediated regulation of EGFR degradation in cancer cells. Not surprisingly, only phosphatase activity at the cell surface was predicted to affect EGFR endocytosis rates. Thus, whereas the activities of phosphatases at the cell surface and in the cell interior may both regulate pEGFR levels directly through receptor dephosphorylation, phosphatase activity at the cell surface may also determine the cellular distribution of EGFR. Overall, our results suggest that the dephosphorylation of EGFR occurs on a time scale which is much shorter than is generally appreciated and that phosphatases may regulate higher level receptor processes such as trafficking in ways which have not been fully explored. Further investigation of these phenomena may reveal new avenues for the rational modulation of EGFR signaling dynamics for therapeutic purposes.

See more of this Session: Modeling Approaches to Examine Fundamental Issues in Life Sciences - I
See more of this Group/Topical: Food, Pharmaceutical & Bioengineering Division

(601g) Direct Reprogramming as a Random Drift in Cell State

Krishanu Saha[1], Jacob Hanna[1], Alexander van Oudenaarden[2] and Rudolf Jaenisch[1], (1)Whitehead Institute for Biomedical Research, Cambridge, MA, (2)Physics, Massachusetts Institute of Technology, Cambridge, MA

Direct reprogramming of somatic cells into induced pluripotent stem cells (iPSCs) can be achieved by overexpression of Oct4, Sox2, Klf4 and c-Myc (OSKM) transcription factors, but quantifying the efficiency and timescales of the crucial events occurring during direct reprogramming has been problematic due to the cellular and genetic heterogeneity of de novo infected somatic cells. To circumvent these issues, we have characterized the reprogramming efficiency and kinetics of over 1000 somatic cell derived monoclonal populations similarly expressing OSKM over an extended period of time. Upon OSKM induction, most if not all monocytes and lineage-committed B cells harboring genetic rearrangements have the potential to generate iPSCs albeit with very different latencies. The observed dynamics are consistent with a continuous stochastic process, in which the conversion from a somatic cell to an iPSC upon continued growth and OSKM expression can be thought of as a drift in cell state. This cell state can be defined by a gene expression or epigenetic pattern, whose fluctuations drive the conversion to an iPSC. Such fluctuations are likely due to cell-to-cell extrinsic heterogeneity or the inherent stochastic nature of gene expression or regulatory signaling processes. Computational simulations and quantitative analyses of these transitions suggest that the number of cell divisions is a key parameter driving epigenetic reprogramming to pluripotency.

See more of this Session: Modeling Approaches to Examine Fundamental Issues in Life Sciences - I
See more of this Group/Topical: Food, Pharmaceutical & Bioengineering Division

(667a) A Deuterium Labeling Strategy for Untargeted Metabolomics Analyses

Curt R. Fischer[1], Paul Wilmes[1], Benjamin P. Bowen[2], Trent Northen[2] and Jillian F. Banfield[3], (1) University of California at Berkeley, Berkeley, CA, (2)Life Sciences Division, Lawrence Berkeley National Laboratory, Berkeley, CA, (3)Earth and Planetary Science, University of California, Berkeley, Berkeley, CA

Untargeted metabolomics analyses are hindered by the difficulty of assigning chemical formulae to unknown features in raw mass spectral data. For that reason, many metabolomics analyses rely on curated databases of authentic standards or known metabolites. Such search-based strategies cannot detect metabolites of unknown structure and function. Here we show that a novel stable isotope labeling strategy – post-extraction deuterium labeling of labile protons – enabled the identification of a family of novel small molecule metabolites that were not present in KEGG, LipidMaps, Metlin, or Metacyc databases. Our stable isotope labeling technique can be applied after metabolite extraction, making it amenable to use even in systems where traditional stable isotope labeling strategies are not possible.

The method was applied to a chemoautotrophic mixed-species microbial biofilm and enabled the identification of a family of highly abundant, unusual methylated phosphatidylethanolamine lipids. Unusual methylation patterns suggest that C1 metabolic pathways play an important role in the biosynthesis of this lipid family. Metabolomic and proteomic analyses of a suite of natural biofilm samples revealed intriguing correlations between these lipids and several rapidly evolving proteins of unknown function. Untargeted, stable isotope labeling-based metabolomic analyses thus have the potential to identify new, functionally important small molecules in microbial systems.

See more of this Session: Proteomics & Metabolomic Approaches to Systems Biology
See more of this Group/Topical: Topical A: Systems Biology

(667b) Study of Mixotrophic and Photoheterotrophic Metabolisms in Cyanothece Sp. ATCC 51142 Via Isotopomer-Assisted Metabolite Analysis

Yinjie Tang[1], Bert Berla[1], Xueyang Feng[1], Bing Wu[1] and Himadri Pakrasi[2], (1)Department of Energy, Environmental and Chemical Engineering, Washington University, St Louis, MO, (2)Department of Energy, Environmental, and Chemical Engineering, Washington University, St Louis, MO

The unicellular diazotrophic cyanobacterium, Cyanothece sp. ATCC 51142 (Cyanothece 51142), is able to grow aerobically in nitrogen-fixing conditions under continuous illumination. This study investigated the impacts of carbon and nitrogen sources on Cyanothece 51142 metabolism via 13C-assisted metabolite analysis and biochemical measurements. Cyanothece 51142 was cultivated using 13C-labeled glycerol, glucose or pyruvate, and the resulting patterns of derivatized metabolites were determined by gas chromatography-mass spectrometry (GC-MS) to track the active pathways. The isotopomer data were then integrated with transcription and enzyme activity measurement to link Cyanothece 51142 physiology with functional genes in the central metabolism. Under continuous light and nitrogen-fixing conditions, we find glycerol addition promoted aerobic catabolism and nitrogenase-dependent hydrogen production (up to 25 μmol H2/mg chlorophyll/hr), but strongly reduced phototrophic CO2 utilization. Under nitrogen-sufficient conditions, Cyanothece 51142 was able to metabolize glycerol photoheterotrophically without measurable CO2 fixation, while its activity of light dependent reaction was not reduced. Isotopomer analysis also detected that Cyanothece 51142 was able to fix CO2 via anaplerotic pathways, and to uptake glucose and pyruvate for mixotrophic biomass synthesis. Based on the unique isoleucine labeling pattern, we discovered that Cyanothece 51142 employs an atypical citramalate pathway for isoleucine synthesis, with pyruvate and acetyl-CoA as precursors. This study demonstrates the merits of combining 13C-assisted metabolite analysis, enzyme assays, and metabolite detection not only to understand cellular metabolisms, but also to discover novel enzyme activities.

See more of this Session: Proteomics & Metabolomic Approaches to Systems Biology
See more of this Group/Topical: Topical A: Systems Biology

(667c) Quantitative Analysis of the 3T3-L1 Adipocyte Proteome During Hypertrophic Enlargement and Its Integration with Metabolic Flux Models

Colby Moya[1], Billy W. Newton[2], Stephanie Cologna[3], Hai Shi[4], William Russell[3], Kyongbum Lee[4] and Arul Jayaraman[1], (1)Chemical Engineering, Texas A& M University, College station, TX, (2) Chemical Engineering, Texas A&M University, College Station, TX, (3)Chemistry, Texas A& M University, College station, TX, (4)Chemical and Biological Engineering, Tufts University, Medford, MA

The increase in adipose tissue mass arises through either an increase in cell number or an increase in the size of individual adipocytes. The latter, termed hypertrophy, occurs in part due to progressive lipid loading and triglyceride accumulation in adipocytes. Hypertrophic adipocytes produce the highest levels of pro-inflammatory molecules and reactive oxygen species (ROS), and not surprisingly, hypertrophy plays an important role in obesity and associated metabolic disorders. Since mitochondria is the site for major metabolic processes such as the TCA cycle and fatty acid oxidation that govern the extent of triglyceride accumulation, and the primary site of ROS generation, we first quantitatively investigated changes in the adipocyte mitochondrial proteome during different stages of differentiation and lipid loading-driven enlargement. Mitochondrial proteins from undifferentiated 3T3-l1 preadipocytes and adipocytes at different stages of lipid accumulation (days 4 through 18) were isolated, digested, and labeled using the iTRAQ 8-plex kit. The labeled peptides were fractionated using an in-house developed liquid phase isoelectric fractionation system (MSWIFT) to increase the depth of proteome coverage, and analyzed using LC-MS/MS. Our data show that the majority of enzymes and transporters involved in the TCA cycle, fatty acid oxidation, and ATP synthesis were up-regulated upon differentiation, and maintain or increase these levels during hypertrophy. Our results also suggest that cultured adipocytes enter a state of metabolic-overdrive where increased flux through the TCA cycle and increased fatty acid oxidation occur simultaneously. The observed increase in ROS was likely compensated by a concurrent increase in anti-oxidant and detoxification enzymes. Since the effects of mitochondrial oxidative stress also impact the cellular enzymatic machinery in the cytosol, we also profiled changes in the cytosolic proteome. A targeted proteomic approach was used wherein enzymes involved in pathways identified as being important using a metabolic flux model (e.g., glycolysis, palmitate biosynthesis) were quantitatively profiled using mass spectrometry. The combination of mitochondrial and cytosolic protemic profiling with metabolic flux models provides a systems-level understanding of processes during hypertrophic enlargement and lipid-loading in adipocytes.

See more of this Session: Proteomics & Metabolomic Approaches to Systems Biology
See more of this Group/Topical: Topical A: Systems Biology

(667d) Carbohydrate Circle and Genetic Information Flow Studies of Escherichia Coli Reveal the Molecular Mechanisms of Resistance to Balofloxacin

Xuan-Xian Peng[1], Jian-Yi Pan[2], Hong-Kai Wu[1], Hui Li[1], Xiang Liu[1], Jun-Xia Gao[1] and San-Ying Wang V[2], (1)School of Life Sciences, Sun Yat-sen University, Guangzhou, China, (2)School of Life Sciences, Xiamen University, Xiamen, China

Protein-protein interactions are pivotal in many biological processes and essential for a global understanding of cell functions. In the present study, we combined 2-D native/SDS-PAGE-based proteomics with Co-immunoprecipitation, far-Western blotting, His-tag affinity purification and functional analysis to establish a reference map of E. coli cytosolic complexome. The reference map included 23 distinct heteromeric and 22 homomeric protein complexes. These complexes mainly contributed to glycolysis/gluconeogenesis, information processing and cellular processes. Out of these complexes, 8 heteromeric and 1 homomeric protein complexes were responsible for baloxin resistance. Functions of these complexes in response to the resistance were categorized into information streams, monosaccharide metabolism, response to stimulus and amino acid metabolic process. Out of them, these complexes contributing to information streams and response to stimulus played a key role in the resistance in comparison with those in response to ceftriaxone, which worked for monosaccharide metabolic process. The results renew the understanding of the acquired quinolone-resistant mechanism that is attributed to GyrA-GyrB complex, indicating DNA metabolic process and translational elongation of the information streams and response to stimulus are also responsible for the resistance. These findings highlight the study of antibiotic-resistant mechanisms.

See more of this Session: Proteomics & Metabolomic Approaches to Systems Biology
See more of this Group/Topical: Topical A: Systems Biology

(667e) Short and Long-Term Changes in Circulatory Protein and Cytokine Profiles Following Burn and CLP Treatments

Mehmet A. Orman, Department of Chemical and Biochemical Engineering, Rutgers University, Piscataway, NJ, Ioannis P. Androulakis, Biomedical Engineering, Chemical & Biochemical Engineering, Rutgers University, Piscataway, NJ, Marianthi Ierapetritou, Chemical and BioChemical Engineering, Rutgers University, Piscataway, NJ and Francois Berthiaume, Department of Biomedical Engineering, Rutgers University, Piscataway, NJ

Burns and trauma lead to a hypermetabolic state characterized by an accelerated breakdown of skeletal muscle protein in the patient which leads to muscle wasting and reduction of lean body Since burn patients in general experience a breakdown in gut barrier function that causes bacterial translocation into the body, patients that develop septic complications exhibit a prolonged hypermetabolic state which is triggered by a systematic inflammatory response. This inflammatory response is mediated by circulatory cytokines or proteins, which in turn might affect the signal transduction in liver resulting in changes in gene expression profiles, which plays an important role in hypermetabolic state. Although some inflammatory cytokines expressed in liver or circulated in the blood, such as IL1-â, IL-2, IL-6, IL-10 and TNF-á, have been extensively studied following thermal injuries (Gauglitz et al. 2008; Klein et al. 2003; Summer et al. 2008), more comprehensive analysis given the dynamic behaviors of all possible inflammatory proteins after burn injury or sepsis is required.

The purpose of this study is to characterize concentration changes in circulatory proteins or cytokines (Eotaxin, G-CSF, GM-CSF, GRO/KC, IFN-ã, IL-10, IL-12 (p70), IL-13, IL-17, IL-18, IL-1á, IL-1â, IL-2, IL-4, IL-5, IL-6, IP-10, Leptin, MCP-1, MIP-1á, RANTES, TNF-á, VEGF) at different time points after burn injury and/or superimposed infection to identify initial (0.5 h, 1h, 1.5h, 2), short term (2 h, 4 h, 8 h, 16h, 24h) and long term inflammatory response (1 day, 2 day, 3 day, 4 day, 7 day, 10 day). This study is conducted using rat models of burn injury and infection. A systemic hypermetabolic response was induced in rats by applying a 20 % total body surface area burn followed by (2 days later) cecal ligation and puncture (CLP) to produce sepsis. Control groups are treated identically except that they are given a sham-burn using 37 C water or sham-CLP, the survival surgical procedure without any puncture in cecum.

An extensive description of circulating protein or cytokine profiles will give the key aspects of inflammation such as pro-inflammation, anti-inflammation, and post-inflammation. This also provides an in-slico model of acute inflammation in the liver that combines the extracellular signals, i.e. circulatory proteins and cytokines, with essential transcriptional responses obtained from microarray analysis.

References

Gauglitz GG, Song J, Herndon DN, Finnerty CC, Boehning D, Barral JM, Jeschke MG. 2008. Characterization of the Inflammatory Response During Acute and Post-Acute Phases after Severe Burn. Shock 30(5):503-507.

Klein D, Einspanier R, Bolder U, Jeschke MG. 2003. Differences in the hepatic signal transcription pathway and cytokine expression between thermal injury and sepsis. Shock 20(6):536-543.

Summer GJ, Romero-Sandoval EA, Bogen O, Dina OA, Khasar SG, Levine JD. 2008. Proinflammatory cytokines mediating burn-injury pain. Pain 135(1-2):98-107.

See more of this Session: Proteomics & Metabolomic Approaches to Systems Biology
See more of this Group/Topical: Topical A: Systems Biology

(667f) Identifying Secreted Biomarkers for Immune Evasion in Cellular Models of Cancer

Yogesh Kulkarni and David J Klinke, Chemical Engineering, West Virginia University, Morgantown, WV

Clinical management of cancer has been revolutionized following the introduction of molecular targeted therapies that promise both specificity and efficacy in treatment. Despite the initial clinical success of these drugs, the emergence of resistance is an increasing, and poorly understood problem. Acquired and de novo resistance to molecular targeted therapy is the manifestation of a complicated biochemical dialogue between the immune system, host microenvironment, and tumor cells that scientists are just beginning to unravel. One potential approach is to characterize the secretome of cancer cells, which consists of proteins that are secreted, shed from the cell surface, and intracellular proteins released into the supernatant due to cell lysis, apoptosis and necrosis. These proteins may contribute to escape from immunosurveillance and end up in the bloodstream thereby serving as potential noninvasive biomarkers. The objective of this study was to identify secreted proteins from two cellular models of breast cancer that may interfere with the immune response of the host. The secretome was collected from two cellular models of breast cancer, BT474 (HER2+/ER+) and SKBR3 (HER2+/ER-); separated using two dimensional electrophoresis (2DE); and identified using MALDI-TOF mass spectrometry. The proteins related to natural cell death, which have no contribution towards the breast cancer secretome, were subtracted from the sum of all proteins released by the cell culture. The impact of the identified proteins on immunosurveillance was characterized using pathway analysis and protein-network enrichment.

See more of this Session: Proteomics & Metabolomic Approaches to Systems Biology
See more of this Group/Topical: Topical A: Systems Biology

AUTHOR INDEX

AUTHOR INDEX

AUTHOR INDEX